Beating Hunger,
The Chivi Experience

T0266789

Beating Hunger, The Chivi Experience

A COMMUNITY-BASED APPROACH TO FOOD SECURITY IN ZIMBABWE

Kuda Murwira, Helen Wedgwood,
Cathy Watson and Everjoice J. Win,
with Clare Tawney

Foreword by Ian Scoones

Intermediate Technology Publications
2000

Practical Action Publishing Ltd
25 Albert Street, Rugby, CV21 2SD, Warwickshire, UK
www.practicalactionpublishing.com

© Intermediate Technology Publications 2000

First published 2000\Digitised 2013

ISBN 10: 1 85339 524 2
ISBN 13 Paperback: 9781853395246
ISBN Library Ebook: 9781780442624
Book DOI: http://dx.doi.org/10.3362/9781780442624

Since 1974, Practical Action Publishing has published and disseminated
books and information in support of international development work
throughout the world. Practical Action Publishing is a trading name
of Practical Action Publishing Ltd (Company Reg. No. 1159018), the
wholly owned publishing company of Practical Action. Practical Action
Publishing trades only in support of its parent charity objectives and any
profits are covenanted back to Practical Action (Charity Reg. No. 247257,
Group VAT Registration No. 880 9924 76).

Contents

Abbreviations

Agritex	Agricultural and Technical Extension Services Department
CBO(s)	Community-based organization(s)
Contill	Conservation Tillage Programme
DFID	Department for International Development of the UK Government (formerly ODA)
DRSS	Department of Research and Specialist Services
GDP	Gross Domestic Product
GTZ	Deutsche Gesellschaft fur Technische Zusammenarbeit (Technical Co-operation arm of the German Government)
HIVOS	Dutch acronym for The Netherlands Organization for Human Development
IFAD	International Fund for Agricultural Development
IRDEP	Integrated Rural Development Programme
ITDG	Intermediate Technology Development Group (in this book understood to mean the Zimbabwe country office of the Intermediate Technology Development Group)
NGO(s)	Non-governmental organization(s)
NLHA	Native Land Husbandry Act
PEA	Participatory extension approaches
POEMS	ParticipatOry Evaluation-oriented Monitoring System
PTD	Participatory technology development
SADC	Southern African Development Community
SAFIRE	Southern Africa Fund for Indigenous Resources
VCWs	Village Community Workers
VIDCO	Village Development Committee
WADCO	Ward Development Committee
ZFU	Zimbabwe Farmers' Union

Acknowledgements

The authors would like to express their gratitude for the immense contributions made to the Chivi Food Security Project by a range of stakeholders. Contributions to the project's success took various forms and were made at different stages of the work. While it is not possible to mention each and every one by name, there are those few worth special mention.

It was not by coincidence that this project became a success story. A lot of energy was spent visioning and programming the work, building upon various lessons and experiences gained internationally over many years of practical work. At the centre of this planning process, working with the authors, were Ebbie Dengu (Regional Director, ITDG Southern Africa), Patrick Mulvany (formerly Agriculture and Fisheries Programme manager, ITDG UK), Adrian Cullis (formerly Dryland Farmers programme manager, Kenya and UK), Dr Ian Scoones (member of the panel to the ITDG, Food Security Projects team), Dr Ian Goldman, and Rodger Mpande (an Agriculture and Rural Development Consultant based in Zimbabwe).

The ideas of the project would not have been able to 'hatch' without the support of the key strategic partner institutions and their staff. The implementation of the project was made possible through the support of the following:

- Agritex in Chivi District (the late Noel Mapepa, L. Makonyere and B. Butaumocho)
- Agritex in Masvingo Province (Munashe Shumba, Osmond Mugweni and Ernest Dando)
- Agritex National Office (Marcus Hakutangwi and Ernest Dando after his promotion from Masvingo)
- DRSS (Isiah Mharapara, Head of Lowveld Research Station, and Dr Phibeon Nyamudeza)
- Mr Silas Hungwe, President of Zimbabwe Farmers' Union
- Zephaniah Phiri Maseko, an innovative farmer based in Zvishavane District, who acted as a source of inspiration to the Ward 21 community.

The offices of the District Administrator and of the Chivi Rural District Council, together with all the administrative units that come under them, played important roles which enabled the work to succeed.

The willingness and commitment of the Chivi community to partner the project initiators needs to be highlighted because 'one can take the horse to the river but one cannot make it drink.' The dedication of the project field staff, Millie Vela, Patricia Mushayandebvu and Monica Nyakuwa should not be taken for granted.

It would not have been possible to scale up the work without linkages to like-minded institutions and programmes, such as the GTZ–Contill and IRDEP Projects, Silveira House, Farming Systems Research Project, Community Technology Development Trust (COMMUTECH), Vredeseilanden Coopibo (VECO), GTZ–SADC Food Security Project, International Crops Research Institute for the Semi-Arid Tropics (Sorghum and Millet Improvement Programme) (ICRISAT(SMIP)), Zvishavane Water Projects, Dabane Trust, Fambidzanai Training Centre, and Makoholi and Chiredzi Research Stations. Many of these institutions contributed to the implementation of the participatory technology development (PTD) process.

A lot of work was done to raise the level of public awareness of the project and its experiences. We particularly acknowledge the following for their sterling efforts in profiling the work and communicating lessons and experiences from the project: Dr Jurgen Hagmann (formerly with GTZ-Contill), Farai J. Samhungu (formerly ITDG Southern Africa), Michael Farelly (Marketing Head), Edward Chuma (formerly with GTZ–Contill), Mike Connolly (GTZ–Organizational Development at Agritex), John Wilson, PELUM, Cecilia Manyame (Consultant), Marcus Hakutangwi (Chief Agritex Training Officer), and Willy Critchley (Centre for Development Cooperation Services, Free University, Amsterdam).

Last but not least, we would like to commend the funders who supported the work from its initiation to its conclusion for taking the risk of investing in a path led by the Chivi farmers. They include Comic Relief, the European Union (formerly EEC), DFID (formerly the ODA Joint Funding Scheme), and HIVOS.

Kuda Murwira, Zimbabwe 2000

Foreword

There is a growing consensus that top-down approaches to technology transfer and agricultural development have failed. These failures have been particularly apparent in those dry and marginal areas of the world where most poor people live. An alternative perspective is beginning to emerge which puts farmers first, is centred on participatory approaches and is based on locally available technologies and resources. Much of the discussion about participatory approaches to agricultural research and extension remains, however, at the level of rhetoric. The wave of enthusiasm for participation amongst donors in recent years has led to some small islands of success, but also to many failures. A big question arises: what needs to be done to encourage a more bottom-up, participatory approach to technology development in marginal areas that has a wider and more sustainable impact?

This book reports on ten years of experience and lesson learning from Chivi District in southern Zimbabwe. Contained in the following pages are some valuable insights of wide relevance to all those working for poverty reduction and improved livelihoods in marginal areas. Of course, in some respects the experiences reported here have been highly particular, the result of contingent circumstance and specific context. However, in striving to make real the key principles of participation of community members in decision making and planning, institutional strengthening, unlocking local skills and knowledge and participatory technology development some important, wider insights emerge. So what generic lessons have been learned? I want to emphasize three linked points. First, participation is not easy; second, approaches based on agricultural technology must be set within understandings of the livelihood setting; and third, scaling up, broader institutionalization and engagement with the policy process are essential.

Encouraging a participatory approach presents many challenges, particularly in settings where local expectations of external intervention are based on top-down command and control approaches. Quick-fix, method-based solutions are clearly inadequate. While the use of a range of different methods for soliciting information and encouraging engaged debate are clearly important, these must be embedded in a strategy for broader social change. During the 1990s the Chivi project used an innovative variety of methods from the participatory rural appraisal toolbox in combination with a continued investment in Training for Transformation approaches. Building individual and collective confidence in the possibilities for change are seen to be an important prerequisite for a participatory approach. As the Chivi experience shows, this takes time, patience and skilful facilitation.

Technology development occurs within broader social, economic and political settings. Such contexts matter. While Chivi is certainly marginal, it is not an isolated, timeless backwater disconnected from the rest of the world. Connections through markets, migrancy, social networks and histories of engagement with the state all impinge on local livelihoods and processes of technology change. External shocks and stresses also have an impact. The initial phases of the Chivi project were dramatically affected by the impact of drought in the early 1990s and the implementation of the government's structural adjustment programme; attempting to construct a separate rural idyll through a project intervention is thus clearly folly. While drawing on local skills, knowledge and resources is an important starting point, as the Chivi experience has shown, this is not enough. Sustainable options draw on multiple sources of innovation – from other farmers, as well as scientists from the local area and beyond. The result is a combination of the new and old, traditional and modern. The key is that the solutions sought are determined by farmers and fit diverse and differentiated livelihood needs.

Too often projects start and finish in their project area. When the support is withdrawn there is little to show. The Chivi project has been different. From the beginning a strategy for institutionalizing the approach within broader structures – from the village to the national level – has been central. One of the starting points of the project has been to build the capacity of local groups to articulate their priorities and, in doing so, to make external service provision (notably through government agricultural research and extension, but also local government and the project itself) more demand-led and accountable. But encouraging village level groups is only the start. To be effective such groups must have some clout. Here the challenges of broader representation come to the fore. How can farmer groups influence research and extension priorities? How can they engage with government policy? How can they guide project priorities? A variety of approaches have been explored in Chivi, including participatory monitoring and evaluation, the 'discomfort' model of training, and facilitating linkages to the national farmers' union. Such strategies for scaling-up present yet further challenges which are only now beginning to be explored in depth. Institutionalizing participatory approaches in large-scale bureaucracies, such as government agricultural research and extension systems, present some rather fundamental challenges for organizational change. Similarly, encouraging participation by farmers in the policy process requires issues of representation and political voice to be addressed.

As the exploration of opportunities continues in Chivi and beyond, these will remain key challenges for the future. But such explorations by ITDG and its partners in Zimbabwe will be based on a very firm foundation, with many important lessons learned over the past decade. For others attempting to set out on this journey, the lessons captured in this book will provide important food for thought.

Ian Scoones, Institute of Development Studies, University of Sussex

1

INTRODUCTION

Participation has been on the lips of development professionals for many years now. People's participation in development requires that the top-down approach be replaced by a process of helping people to articulate their problems, form self-help groups and formulate plans to use their own resources to achieve their objectives. Such methods have been used in projects to rebuild infrastructure in the wake of cyclones, to organize villagers to run their own night schools for adult education, and to form groups to guarantee jointly the repayment of business loans.

'Participation' aims to empower people through joint activities; the alternative style of intervention – imposing projects on passive recipients, or offering them project 'handouts' – is disempowering, and creates dependency on the external agency. In an era when aid is increasingly rationed across ever-expanding areas of need, helping people make the best of their own resources rather than hope for bounty from outside seems to be the only realistic policy.

For poor farmers in the marginal lands around the world the drive for participation in agricultural projects comes with an additional impetus. The goods from outside – in this case the 'green revolution' technologies of high-yielding seed varieties, packaged with fertilizers, herbicides and pesticides – are not only unaffordable, but they are inappropriate. It is now well recognized that introducing high-input agriculture to areas of the world where poor people cultivate difficult soils in adverse climatic conditions is not successful, not because the farmers themselves are ignorant or are unable to apply the new technologies according to the book, but because the technologies are often not suitable for the variable conditions under which the farmers work.

Technologies developed on research stations and promoted by agricultural extension services may yield well in the prosperous lands of, for example, the Punjab or the commercial farming regions of Zimbabwe, but perform erratically in the semi-arid, dryland farming regions of southern India, and the communal lands of Zimbabwe. The result is that farmers are barely able to cover the investment costs of these seed packages and feed their families as well.

Those areas where modern varieties were successful can be termed 'green revolution' regions: where rainfall is reliable, or irrigation possible. Marginal lands which are distributed all over the Third World, and cover much of sub-Saharan Africa, have been characterized as complex, diverse and risk-prone (Chambers 1989): here agriculture is rain-fed, and droughts are a regular occurrence.

This book describes the Chivi project, a pilot study which explored alternative ways of working with smallholder farmers to develop technological options appropriate for such conditions. The first part of this chapter considers farmer innovation and technology development, and contrasts this with the failure of agricultural extension services to deliver relevant technologies for marginal conditions. Two themes are then explored: participation – but by whom and to what extent? – and the sustainability of livelihoods and institutions.

Farmers as experimenters

Poor farmers in marginal lands traditionally plant a range of crops and vegetables, often intercropping, as well as preserving trees in their fields for their fruit and firewood. To the eyes of most commercial farmers, their farms appear an untidy mixture of crops, with tiny plots and low yields. Agricultural scientists and social scientists increasingly recognize, however, that what poor farmers have always used – their own varieties and traditional farming techniques – are well adapted to their environment, and are more appropriate than the monocultures that research stations and extension services tend to promote. From all over the world, practices previously dismissed as unscientific and backward are now recognized to have a rational basis, and to be the result of informal innovation and experimentation on the part of often uneducated farmers.

There is now plenty of evidence that farmers are skilled at selecting seed varieties to suit the varying agronomic conditions on their farms, and plant a range of varieties to suit different soils and water availability within their landholding. Mende farmers in Sierra Leone, for example, select shorter- or longer-season rice varieties to plant in marshy areas, in water-retentive soils in valley floors, in free-draining upland areas, and in water courses – and thus avoid the need to irrigate (Almekinders 1999). This adaptation of the germ plasm to suit the conditions contrasts with the conventional approach of adapting the conditions (for example, applying irrigation) to suit the requirements of modern varieties. Farmers identify favourable characteristics in other farmers' varieties and combine these with their own varieties, selecting the plants with the most promising characteristics and multiplying up the seed. While formal research tends to concentrate on yield alone, farmers' selection criteria are more complex and include taste, texture and cooking properties, suitability for intercropping, pest resistance, ability to grow in the shade of trees, length of duration, reliability of yield during drought years, and so on.

All this does not mean that farmers are producing enough to feed their families: regular food relief in these regions is evidence to the contrary. Nor does it suggest that farmers have all that they need in terms of varieties and technologies at their disposal, and that agricultural research stations or extension services have nothing to offer them. Farmers are interested in formal variety trials, as was demonstrated by the case of the Mahsuri rice, a variety that 'escaped' from a research station in India via a farm labourer, was multiplied and spread informally among farmers with whom it was popular, and was only officially released by the government some time later as a result of popular demand (Maurya 1989). It is also a mistake to label modern varieties as 'bad' and traditional varieties 'good' for marginal farmers – they themselves are usually interested in trying out the new alongside their native varieties, incorporating a new variety that has performed well in their fields within their own germplasm banks along with several of their favoured native varieties; indeed they probably do not see a great dichotomy between new and native varieties (Rhoades 1989).

What is becoming clearer is that, when farmers adopt new technologies, they do so incrementally, trying out different elements that seem most promising rather than investing resources in the whole new package. This can be seen in the example of the introduction of diffused-light storage shelters for seed potatoes: there was apparently a low level of adoption of this technology until it was discovered that, although the farmers had not gone to the expense of building new shelters, they had incorporated the principle of diffused light into a variety of different storage situations (ibid.). Similarly, rather than unreservedly adopting a new variety, many farmers buy a small quantity of seed in order to combine particular genetic characteristics of the new variety with their own seed, and so get the best of the new while preserving the adaptability to adverse conditions of the local variety. What farmers want, it seems, is not an all-or-nothing package, but a 'basket of technologies' from which to chose what is most appropriate for them.

What then is the role for 'outsiders' – whether agricultural researchers, extension workers or non-governmental organization (NGO) project workers – in farmers' experimentation? Rather than instructors in improved methods, they need to act as facilitators of experimentation, convenors of meetings, suppliers of genetic material, or even as travel agents to arrange visits between farms. This is the role played by the project officer in the Chivi project; and chapters 7 and 8 describe how exposure visits to research stations were arranged, farmers discussed which techniques they wanted to try out, and trials were carried out and evaluated. Taking farmers to view a range of technical options placed the onus upon the farmers themselves to choose what they considered most suitable, and left them free to drop whatever they decided had not performed well in the trials.

The gap between farmers and extension workers

The story of the agricultural extension service in Zimbabwe provides an essential background to our understanding of the Chivi project. The situation of the farmers in the communal areas of Zimbabwe, outlined in chapter 2, derives from a history where legislation deprived Africans of the best areas and resources, and forced them to farm the poorest land. Men were encouraged to work away from their homes, leaving the farms to be run single-handed by women. Agricultural extension messages in the colonial era were designed with the conditions of the wetter, commercial agricultural areas in mind, and as well as being detrimental in some ways, they were enforced upon the African farmers by legislation: those who refused to implement the prescribed measures could be punished.

In Zimbabwe the differentiation between rich and poor farmers, and the capture of the best farming resources by the rich, was driven not just by economic forces but by legislation based on race. The lack of understanding between the extension service and marginal farmers, which around the world is often caused by differences in education, background (often urban vs. rural), life expectations (career development vs. survivalist farming), was here exacerbated by the divide of race and culture. It is small wonder that the techniques adapted for such complex, diverse and risk-prone areas were overlooked rather than investigated.

Chapter 2 describes how this imposition of unpopular measures was halted at independence, and a commitment made to grassroots development. The agricultural extension service has operated for the most part on a 'transfer of technology' model, however, with extension messages being directed mostly at the farming élite of master farmers: richer farmers who had undergone some training and who had gained a diploma. The hope was that other farmers would copy the methods demonstrated by the master farmers; the reality was that these technologies were considered unsuitable by poorer farmers, and the anticipated 'trickle down' did not take place. Most farmers received little attention from the hard-pressed agricultural extension workers and those who carried out their own experimentation or employed indigenous methods were more likely to be secretive about what they had done than share their expertise.

Such experiences are not unique to Zimbabwe but also apply to many agricultural extension services around the world. It is estimated that 80 per cent of all agricultural extension 'messages' in India are not taken up by farmers – and it is probably more than this for advice for dryland farming (Chambers 1989). Carney (1999, quoting World Bank 1994) writes:

> Notable successes such as the yield increases of the Green Revolution, have been offset by notable failures such as collapsed rural credit schemes and research and extension systems which remain dysfunctional despite enormous investment over the years. For example, a 1994 review of World Bank extension project found that 90 per cent

experienced recurrent cost funding problems and 70 per cent were probably not sustainable. Since the Bank committed over $1.4bn in new loans to extension during the period 1987–93 the magnitude of this underperformance problem was significant.

ITDG's background in food security

The Agriculture and Fisheries Sector, ITDG's group that managed food security projects internationally in the 1980s and early 1990s, defined technology as 'skills and knowledge used by people to produce goods and services'. Appropriateness was determined more by social, economic and institutional 'fit' than by mechanical or technical assessment. It was underpinned by the maxim expressed by Fritz Schumacher, founder of ITDG: 'Find out what people are doing, then help them to do it better'. When work was initiated in Zimbabwe in the late 1980s it was within this internationally determined framework.

ITDG's Agriculture and Fisheries programme developed a people-centred approach in the 1980s. Pioneering work was done in Turkana in a water-harvesting project, which focused on strengthening local Turkana institutions, adekars, so that they decided the pace and nature of any interventions to sustain sorghum and livestock production and increase food security. Within ITDG, the Chivi project was seen as a way of testing some of the lessons learnt from Turkana: the importance of local organizations, of dialogue with the community, and of continuity. Other areas in ITDG were taking a similar approach, for example, the Maasai Housing Project in the Kajiado District of Kenya was using participatory technology development (PTD) as a means of increasing access to decent and affordable shelter.

In Zimbabwe, the context was one of the colonial heritage of food insecurity of the highly-populated and marginal agricultural 'communal areas' in agroecological zones IV and V. In these areas, food-aid disbursements were a routine drain on national resources. ITDG Zimbabwe placed a high priority on looking for alternatives for improving local food security in these areas, through appropriate and replicable interventions.

After a long search in co-operation with national, provincial and district government authorities and non-government institutions and discussions with local development committees, the way of identifying the location for the initial pilot work was agreed. Chapter 3 describes how Chivi District was eventually selected as being typical of the zone, and the district authorities with local ward committees finally decided that the work should begin in Ward 21.

At the outset there were many pressures brought to bear on ITDG by government and donors (even ITDG's UK management) to deliver 'technical solutions'. Reason prevailed, however, and the slower, more thorough process described in this book was developed.

Facilitating participation

The original objectives of the Chivi project were:

- To increase household food security through improved agricultural production;
- To strengthen local institutions and enable poor farmers (men and women) to articulate their priorities and control productive resources;
- To influence government agricultural policies to be more responsive to the concerns and circumstances of poor farmers.

The Chivi project built on the growing recognition within ITDG that technology development is a social process and that projects should be 'client-led' rather than 'technology-led'. Given this, it may seem contradictory that objectives for the project were decided upon before the people concerned could be consulted, and before their needs could be elicited. In addition, the choice of the project area itself – using criteria to select a resource-poor area where food security was an issue – was made without waiting for an invitation from the people. Once the project was well known in the area, villages did of course have the option of refusing to participate. This happened in one village, where for political reasons the *sabhuku* forbade farmers' club and gardening group members to join in.

Working within carefully chosen boundaries does not necessarily mean that control is taken away from local people, however; it simply means that the activities chosen and planned by the people are concentrated in one area of need – food security – and are less likely to become dissipated and ultimately unsuccessful. Encouraging people to voice their needs often leads to a 'wish list' that must be met with refusal and disappointment, and raising expectations that irrigation tanks may be dug, schools built and land reallocated would be to take on the role of fairy godmother, and may ultimately be disempowering. On the other hand, improvements achieved through building on people's own skills and ingenuity are both sustainable and empowering. Chapters 4 and 5 describe how the project elicited people's needs, helped them to prioritize their objectives and encouraged planning to meet these objectives.

Another aspect of participation is: participation by whom? Many research projects in the 1980s had stressed the participation of individual farmers in technology development, but the Chivi project went beyond this in stressing the importance of the participation of all farmers through their involvement in local institutions, so as to increase the food security of all. Local groups were seen to be the best way to encourage some of the poorest people to participate, and indeed there is evidence presented in chapter 9 that 80 per cent of the lowest wealth rank in the project villages were benefiting from the project.

Working with local groups also served to ensure the participation of women. Gender should be inherent in a participatory approach to development, but it is not automatically addressed without specific efforts (Frischmuth 1997). In Chivi, alongside the farmers' groups, that were mainly the preserve of men engaged in field operations, garden groups with a membership of women were selected as the institutions most suitable for achieving food security aims. Once this was decided upon, it was much easier to ensure equal participation of women in meetings, on training visits, and in leadership roles, simply because the project was working with each group equally, and without needing to confront head-on cultural norms about women's position in the background of village life. Even success is not without its problems, however: after the women achieved a bumper harvest of groundnuts one year, some men expressed their interest in taking over groundnut production (see chapter 10).

Participation must also not be confused with listening to the directives of the last or loudest voice. The facilitator, far from taking a passive role, has to be able to elicit the opinions of the quietest participants of the group, without obviously dismissing the loudest. There are many techniques under the umbrella of participatory rural appraisal – some of which are described in chapter 4 – which help reveal the alternative viewpoints of different people. One element of the programme proved indispensable for boosting the confidence of quieter members and challenging the position of more forward villagers: Training for Transformation courses played a crucial role in making the farmers' groups and garden groups more democratic and inclusive (see chapter 5). Even agricultural extension workers who had been sceptical of the need for Training for Transformation – one commenting 'I had worked in the field for many years: why did I need more training?' – became convinced that what they had learned on the course about encouraging self-reliance among the farmers and treating them with greater respect was a necessary part of successful extension work.

One further comment must be made about the project's objective of food security. Current development thinking on sustainable livelihoods emphasizes that rural people do not rely on agriculture alone for their survival, and that projects should address their wider concerns as, say, handicraft producers or fishermen (Carney 1999). The single focus on food security in the Chivi project may seem to contradict this. Nevertheless, while the focus of the project was 'food security', the team worked to a fairly open agenda, and the scope of the project was broad, for example following up people's concerns for fencing. Even when needs were voiced that lay outside the area of food security, they were not necessarily set aside: as the project expanded out of Chivi into Nyanga District, other needs such as those of the gold panners were dealt with by project staff. This reflects the recent move within ITDG for projects to include the strengthening of non-agricultural aspects of local people's livelihoods – in other words for food security to become livelihood security. There are strong arguments nonetheless for concentrating on food security and making an impact, particularly when the success of the project relied on people observing the success of other farmers and choosing to copy their methods and participate themselves.

It is clear, then, that although the project aimed to promote planning and decision making by local people, it had its own objectives and strategies for achieving them. In the case of gender participation and involvement of the poorest, these were not always shared by all the villagers. As one observer put it: 'There was what one might call "facipulation": a mixture of manipulation and facilitation where the NGO used a strong approach with a singular focus so that activities of the project were in line with the philosophy of the NGO. The programme then invited the target group along that path. It was not a case of the NGO being invited along the path chosen by the target groups' (Mbetu 1997).

Building sustainable institutions

Sustainability is a repeated theme in this book. The word has been overworked in the last few years, but no apology is made for using it again since the sustainability of the farming techniques and the sustainability of the local institutions involved are key to the success and innovation of the project. For marginal farmers all over the world, the greatest fear is not so much low yields as uncertain yields, just as poverty is experienced not simply as low income but as a sudden need for cash when there is no way to meet that need. In other words, poverty is seen as vulnerability to shocks (for example, drought or pest attacks), and the project based its interventions on building people's resilience to such shocks, or creating 'sustainable livelihoods' (DFID 1999). The soil- and water-conservation measures and the new varieties that were tried out and adapted involved little cash outlay, and were therefore low risk. Intercropping, cultivating a broad range of crops, and staggering planting times, together with a variety of small off-farm business ventures often run by a single household, are all part of a risk-reduction strategy to withstand disaster.

Sustainability – in this case, sustainability of the project's impact, or making a lasting difference – is also a prime reason for working with local institutions. Existing groups have had a life before the project arrived, and are more likely to endure, strengthened, after the project has departed. Creating new groups may be a tempting option in the light of the power struggles or undemocratic organization that may be apparent in established groups; however, it is possible to encourage in existing groups a wider membership which can make more demands on the leadership, and evidence of this happening in the Chivi project is given in chapter 9. What is critical is that the institutions chosen are flexible enough to develop in this way, and to encourage full participation by all village members. The process of selecting institutions is described in chapter 4.

The emphasis on local institutions developed in Chivi has also significantly influenced ITDG's ways of working elsewhere. Seven of the eleven projects in the current programme focus on PTD methodologies and methods to increase agricultural productivity. Many of the local partner institutions, local councils and community-based organizations (CBOs) are formulating local community development plans and using participative methods to monitor and evaluate them. The organization of seed fairs has been successfully handed over from projects to local community groups. Also the traditional livestock healers in Kenya, *wazaidizi* (literally 'helpers' of livestock), and farmer-trainers in Peru, *kamayoq*, have formed their own self-help associations. Chivi should not be seen as a blueprint, however. One of the key lessons emerging from current work is that the transition from project to a strong CBO, although possible, is complex and lengthy. It is necessary to be more realistic in what can be achieved over the lifetime of a project and to recognize diversity in different circumstances.

Institutionalizing farmers' participation

Recognizing the failure to get extension messages across, and aware of the untapped potential of marginal farmers to innovate successfully, research interest is moving on to ways of institutionalizing PTD. This means that participatory approaches should be adopted not just by single research projects but by research institutions and agricultural extension services. This was one of the stated objectives of the Chivi project. It also has a bearing on sustainability, of course: the Chivi project has a limited lifespan, but the support of the extension services will endure.

Institutionalizing participatory approaches is not easy, however. As its name implies, the agricultural extension service is designed to deliver agricultural techniques one way from a central department to farmers, not to learn from farmers. These services, like agricultural research stations, are controlled by bureaucracies, and as such have a tendency to lay down fixed ways of working, with knowledge formalized in reports or scientific publications, and rewards based on centrally agreed objectives. It is difficult for agricultural extension workers to support PDT among farmers: to send farmers' queries or new techniques 'up the line' to their superiors rather than pushing technologies down to farmers, to write reports when activities are varied, and to justify the use of a vehicle when transporting farmers to visit other innovative farmers.

Even research projects that are designed to consult farmers can run into difficulties documenting farmer innovations. For example, at an operational research project in India in which technologies were evolved and formalized at the research station, it was found that during trials on the farmers' fields, modifications were introduced that were often more cost effective. As yet no working arrangement had been devised to formalize those practices which were modified at the village level (Sanghi 1989).

One of the great strengths of the Chivi project is the involvement of the agricultural extension service from the start. It is significant that ITDG's Country Director was formerly a director of Agritex, and Agritex officials were included in the earliest discussions about the design of the project. Local extension workers were involved in all activities in Chivi Ward 21, from introducing the ITDG project officer to the villagers, to attending training sessions and exposure visits.

Almost ten years later, as a result of the influence of the project, the agricultural extension service nationally has adopted participatory extension approaches (PEA) as part of its training for all its staff, and is implementing projects based on the process approach described in chapter 5. Chapter 12 describes these changes, which have now reached the stage where agricultural extension workers are putting into practice participatory extension programmes in several provinces of Zimbabwe, with advisory support from ITDG and GTZ.

It is still early days to see whether Agritex can cope with the changes in orientation demanded by the participatory approach, or whether the initial enthusiasm will be replaced by lip-service to participation, as has happened elsewhere (Chambers 1989). A recent review by Agritex of its performance revealed that extension workers still need to change their attitudes and ways of working with smallholder farmers. In addition, external problems have arisen, such as a cut in the petrol budget reducing staff time spent in the field. Nonetheless, extension workers are managing to take farmers to visit other innovative farmers, and new partnerships are forming between farmers, researchers and extensionists to conduct experimentation. Many signs are encouraging, in particular the enthusiasm apparent in individual extension workers and recorded in these chapters for approaches that obviously work with farmers.

Whether willing but hard-pressed extension workers are able to become more client-led depends to a large extent on farmers putting demands upon them. Whether the farmers of Chivi District Ward 21 continue to plan, experiment and share their results also depends to a large extent on themselves. In the event of budget cuts to extension services or the removal of NGO project staff, in the end farmers have only their own resources to rely on, as they always have, and if the project succeeded it was only to the extent that it boosted farmers' confidence and developed this resourcefulness.

The Chivi project has received a lot of interest both from within ITDG and externally. Key people, who visited the project early on, were sufficiently interested by what they had seen that they also helped to spread the word. This has generated a lot of interest from other NGOs and researchers throughout the world and has contributed to a constant demand for information about the project. We hope that the chronicle of events and the analysis of outcomes presented in this book will help to fulfil some of these information demands, and provide useful ideas for others involved in strengthening the skills of poor farmers around the world.

2
BACKGROUND TO THE PROJECT

The single most important reason why the African countryside has not developed has been the fact that its people have lost their autonomous rights to make and implement decisions concerning their own future. The shape of their future, collectively and individually, has been drawn for them by others, so their failure to identify with it is hardly surprising.

(Cheater 1989:2).

The Chivi Food Security Project was initiated in 1990 by ITDG (Intermediate Technology Development Group's office in Zimbabwe), to explore methods of working with local institutions to increase household food security through more responsive agricultural and other extension services. 'Participatory' development approaches that encourage farmers to analyse their problems and plan their own projects have been around for at least 20 years, nevertheless the existing agricultural extension service in Zimbabwe, although espousing a 'bottom-up' approach, was far from being participatory. Instead, agricultural techniques, designed in research stations often with richer commercial farmers in mind, were adopted at the national level and pushed by agricultural extension workers onto poor farmers in the communal areas. The uptake of such techniques by these farmers is very low.

Around the world top-down extension services make the mistake of delivering inappropriate messages to the poor majority of farmers. The particular problem in Zimbabwe, however, is that the people farming on the poorest communal lands had arrived there not purely as a result of economic pressures, but because of legislation in the colonial era that consigned black people to these areas. This was accompanied by agricultural extension based on the belief that the techniques of the poorer farmers were backward, and should be replaced by the radically different methods of 'modern' farming.

This chapter provides a background to farming in the communal lands in Zimbabwe, and attempts to explain how historical events left poor farmers in the position where they were unable to benefit from farming advice from the government, but also had little belief in the few indigenous techniques they still remembered. It describes the participatory process approach, and suggests why it was so crucial to build farmers' confidence in their skills and in their own resources to organize in order to have lasting effect.

The context of the project: Zimbabwe

Agriculture makes a significant contribution to the Zimbabwean economy, accounting for 13 per cent of GDP (Mutimba 1994). Production is based on three distinct sectors: first, the large-scale commercial farms, which are situated in the high, fertile land in the centre of the country, and produce cash crops, many of which are exported. These were originally white settler farms and many are still owned by white farmers today. Second, small-scale commercial farms are generally owned by black Zimbabweans and are scattered across the country. Third, the communal areas, formerly tribal trust lands under the colonial government, occupy the lower, more marginal parts of the country.

The communal areas are home to approximately six million of Zimbabwe's ten million people and consist of communal grazing areas, interspersed with individual family plots for crop production. They are characterized by poor soils and low and erratic rainfall, and fall largely into land classification regions IV and V, defined as land suitable only for semi-extensive livestock production (ITDG 1995a). Chivi District, where the project on which this book is based lies, is in Masvingo Province in the south of Zimbabwe, in land classification region IV. Only 15 per cent of total output is marketed from the communal areas, compared to 94 per cent from commercial farms (ITDG 1991a).

The government agricultural extension service, Agritex, has a well-developed network of officers and extension workers, covering the whole country. In the communal areas, the ratio of extension workers to households is roughly 1:800 (Mutimba 1994). Extension advice focuses on high-input agriculture for cash crop production, usually taking the form of recommendations developed at national level. Uptake of Agritex recommendations is poor in the communal areas.

A rich African agricultural history

There is plenty of evidence that in the nineteenth century, before the changes that accompanied colonization took place, black farmers had a system of agriculture that was successful enough to allow them to feed themselves and to trade with the surplus. The British moved into the area partly because of the difficulties they were having participating in the lucrative gold-mining industry in South Africa. However, when they realized that the gold deposits they had anticipated finding in what was to become Zimbabwe were much smaller than those in South Africa, they turned to agriculture, which seemed more promising because of the favourable climatic conditions and soils in most parts of the country.

Upon arrival, the settlers found the local people active in food crop production, tobacco for use and trade, as well as trade involving salt, ornaments and copper products. The local people also made implements such as hoes for tilling the land (Manyame 1994). Before colonization the Shona traded with countries as far afield as China, as evidenced by Chinese artefacts found at the Great Zimbabwe Monument. An important trade commodity was salt which was essential for their diet (Beach 1980).

Rodney describes how indigenous people had made advances in agriculture and mining well before the settlers had arrived. He points out that Zimbabwe had produced hydrologists who had diverted countless small steams for irrigation. These streams were 'made to flow around hills in a manner that indicated an awareness of the scientific principles governing the motion of water. On the mining side … the African people had produced prospectors and "geologists" who had clear ideas of where to look for gold and copper in the sub-soil' (1972: 77).

One of the misconceptions about pre-colonial agriculture in Zimbabwe is that it was subsistence; it was not. When the settlers arrived in the country, black farmers were ready to exploit the new market and provided the settlers with grain and livestock for food. Bourdillon argues that in 1903, African (and predominantly Shona) sales of grain and stock amounted to £350 000 (sterling). Subsequent pressure from white farmers brought about a deterioration in black agriculture, first by pushing for legislation which protected white farming against black competition. 'So when we speak of traditional Shona agriculture as "subsistence" we must beware of belittling its economic viability' (Bourdillon 1987: 71).

With regards to food production, small grains such as finger millet, sorghum and pearl millet were the staple crops. Other food crops included groundnuts, pumpkins, melons, cowpeas, sugar beet and maize. The cultivation of these crops and the usage of natural resources such as sources of water (wells, springs and rivers) wildlife and even mining were all closely related to the socio-cultural fabric of the community. Before and near the start of each rainy season, rainmaking ceremonies were carried out prior to tilling the land. During the growing season, there was a sacred day on which the whole community would form a work party to cultivate an individual farmer's field, thus reducing the burden on one family or household. People were not allowed to work on their family fields on that day. If they were not participating in a work party at a community member's field, people did other chores such as gathering firewood, fishing or anything other than working in their own field. Water sources and the harvesting of wild fruit were also governed by traditional rules and values. Transgression of the sacred values was punished. Some of those rules included the following:

- No one was allowed to cultivate land or settle in the catchment area of water sources such as springs;
- Certain trees were not to be used for firewood and trees near permanent water sources were not cut down; and

- Tilling the land immediately after the rains was not allowed; people had to wait for one or two days.

The erosion of indigenous farming methods

These traditional conservation methods were gradually dropped during colonial rule, when local people lost ownership rights and control of their water sources. The social fabric and culture of responsibility were under attack and this resulted in a change of relationships within the community. For example, 'the sacred role of the spirit mediums was transferred to the local chiefs who were only able to act on the instructions of their masters' (Murwira 1995). In some parts of the country this exchange and imposition of roles led to a complete distortion of the people's social system. In the north of the country, for example, the Tonga people were generally matrilineal and women often held the position of chief, but the colonial government appointed male chiefs, based on other criteria such as being a popular traditional medicine practitioner (Manyame 1994). The Tonga social systems were gradually eroded to the extent that the Tonga have become generally patrilineal. Such changes had the effect of deskilling and confusing communities, as people assumed positions and responsibilities for which they were ill equipped.

Before the introduction of the plough in the twentieth century, agriculture was based on livestock and shifting cultivation. Livestock provided meat and milk, and were used for traditional social obligations. Skins were used to make clothing. Livestock were also used for transport and to supply manure, but were not used for draught power (Hagmann and Murwira 1996). The slash-and-burn systems practised in Zambia and Zimbabwe, which allowed the land to lie fallow and thus enhance fertility, were discouraged by the colonial authorities and eventually abandoned by communal farmers. With regard to tillage, communities practised minimum tillage, and the use of axes and hoes minimized soil erosion. Of course there was a smaller population at that time and less pressure on the land, so the threat of soil erosion was not so great. However, a recognition of the appropriateness of indigenous conservation methods, accompanied by an equitable distribution of land, might have minimized the land degradation that eventually occurred.

The introduction of animal draught power and the plough at the turn of the century, and the shift from growing maize as a food crop to a staple, contributed to soil erosion, rill and gully formation, and food insecurity. Specifically, the plough led to the cultivation of more land per household as rural farmers adopted the plough as a faster piece of equipment for tillage. In 1905 the number of ploughs in use by Africans was 440, by 1921 this figure had risen to 16 900, to 53 500 in 1931 and to 133 000 in 1945 (Arrighi and Saul 1973).

The use of the plough was part of a package of agricultural techniques promoted by the American missionary, Alvord, in the 1930s (Page and Page 1991).

These techniques included monocropping, planting in rows, and uprooting trees from the field in order to make ploughing easier. The introduction of maize and maize grinding mills in the 1930s and 1940s contributed to the shift from food crop production to money market production and the erosion of small grains which were suitable for the semi-arid lands to which the Africans had been driven. The introduction of the plough, the removal of woody vegetation from fields and a reduction in fallow periods all contributed to soil erosion and gully formation in most rural areas.

From 1930, erosion and degradation had become so severe that the colonial government promoted highly unpopular conservation measures including de-stocking, through such legislation as the Native Land Husbandry Act of 1951 (NLHA) (Hagmann and Murwira 1996). Conservation measures such as contour ridges, storm-water drains and waterways were meant to stop further land degradation, but this type of conservation layout was not suitable for semi-arid conditions. On the commercial farms, where the rainfall was much higher and the soils heavier, these conservation methods did stop rill erosion by draining excess water off the fields, but this was not necessary or desirable in drier conditions. Despite the different climatic conditions, these measures were imposed in the semi-arid areas where the blacks had been moved to and where water retention systems would have been more appropriate (ibid.). Farmers who practised intercropping were forced to uproot their crops, and others who rejected the conservation methods were punished.

The combined impact of these unsuitable conservation strategies and, since the 1920s, the coercive approach to extension, overriding what were considered to be backward methods, has weakened farmers' confidence in their indigenous knowledge. It is small wonder that individual farmers in the Chivi project were usually secretive about any existing techniques they practised, ascribing any particular success they had to luck, rather than to their knowledge of an indigenous crop protection method (see Box 7.1).

The point here is not that indigenous methods are necessarily the answer for marginal farmers in Zimbabwe today – nobody suggests a return to slash-and-burn agriculture. Methods that worked when the population in these arid regions was much lower, or before people had requirements for cash, may not all be suitable today. Neither are modern farming methods or modern varieties necessarily bad for marginal farmers – the farmers themselves are usually interested, given the chance, in trying them out. What is important, however, is that as new technologies were introduced into these areas, rather than the farmers being encouraged to adapt them and evaluate them alongside their existing methods, they were imposed upon them, and other methods were discouraged.

The removal of resources from African farmers

The position of poor black farmers was a result not only of an administration that believed it knew best about how their land should be farmed. This assumed superiority of educated public sector employees over illiterate farmers occurs in many parts of the world, as does the imposition of agricultural technologies designed for more productive land upon poor people farming in marginal areas. What is specific to colonial Zimbabwe (or Rhodesia as it was then) is the way Africans were forced into the less productive areas, and then prevented from using their labour and resources freely to make the most of their farms.

During the colonial period agricultural loans were made available only to the white settlers, which meant that it was difficult for African farmers to buy inputs for their crops. In 1912, the Land Bank (later the Agricultural Finance Corporation) was set up with a share capital of £250 000 (sterling) to make credit facilities available to persons of European descent only, at 6 per cent interest, payable within ten years.

The problems of the underdevelopment of Zimbabwe's rural areas can be traced to the political economy of the colonial era. The agricultural policies of this era were based on separate development in favour of the white settlers. Various pieces of legislation were put in place to push the Africans off productive lands into natural regions IV and V, characterized by low rainfall and poor soils, and the poverty of the rural areas is a direct result of these policies. In the early days of settlement, for example, the settlers employed various methods of extracting labour (which was crucial for commercial agriculture) from the indigenous people. In 1894 the settlers introduced a hut tax of 10 shillings for every adult male. In 1904, this was replaced by a poll tax of £1 (sterling) 'on each male over sixteen and 10 shillings upon each wife exceeding one'. When the hut tax was introduced, payment in kind was accepted, but it was soon discouraged in order to induce Africans to earn their tax by wage labour (Arrighi and Saul 1973: 194). The introduction of taxes and wage labour increased women's workload as the men left the rural areas to go and work on commercial farms in the towns and cities. The Native Registration Act of 1936, which stipulated that blacks stay out of white-designated areas unless they worked for the white man on whose land they resided, enhanced the 1930 Land Apportionment Act and contributed to the maintenance of wage labour on commercial farms (Gaidzanwa 1988: 2).

Other pieces of legislation that promoted commercial agriculture for the whites and subsistence farming for the blacks included the Land Apportionment Acts of 1930 and 1940 (Holleman 1952: 45). The Land Apportionment Act of 1930 increased pressure on the land when large numbers of Africans were herded into the 'native lands', later renamed Tribal Trust Lands by the Land Apportionment Act, No. 23 of 1962. Through the 1930 Land Apportionment Act, the productive land on which the whites farmed was termed private, while the land reserved for black people was held under traditional tenure and user-rights. The 1951 NLHA put more pressure on land available to black farmers, increasing the movement of blacks from white commercial areas, introducing freehold tenure, and enforcing de-stocking and other unpopular mandatory conservation practices (Rukuni and Eicher 1994). The NLHA and the approach to the problems of land degradation did not slow down the growth of poverty and further degradation, instead the processes of dependency and the erosion of indigenous knowledge were enhanced. Communal farmers became more dependent on extension workers for solutions to issues relating to tillage, pesticides, seeds, seed varieties and drought-resistant crops.

Development policies in post-independence Zimbabwe

Box 2.1   The government embraces a grassroots development
approach

Peoples' participation is a prerequisite for rural development activities. It is therefore necessary that a bottom-up planning approach is employed. The expressed needs of the rural population and grassroots development proposals must be brought up to the district and provincial levels, where they have to be reconciled with the Central Government's views and possibilities.
1983 Zimbabwe Government strategy paper, *Towards the Implementation of a National Rural Development Policy* (Makumbe 1996)

The ruling party, having recently come out of a protracted armed struggle for political independence, was guided by socialist ideology which emphasized popular participation, and called for the active involvement of the majority in rural development. The argument was that through self-development, the people would not only be free of the effects of colonialism, but that they would also be responsible for their own destiny. The struggle for the independence of Zimbabwe had been fuelled by the need for land and the black majority's need to control the economy. At independence the new Zimbabwean Government was under pressure to redress the imbalances of the colonial era and it introduced policies that were meant to benefit the majority. Of note were policy documents such as the Transitional National Development Plan 1982/84–1984/85, the First Five-Year National Development Plan 1986–90, Growth with Equity and the National Land Policy.

During the struggle for independence, the conservation measures enforced under the NLHA were regarded as a symbol of oppression of the white minority, and farmers were encouraged to pull down the conservation ridges, or at least not to maintain them. It then became politically difficult on the part of the

Government of Zimbabwe to reintroduce the 'colonial' conservation measures, and soil- and water-conservation techniques were neglected by the extension service in the 1980s in favour of achieving the commercialization of smallholder agriculture.

With regard to food security, the Zimbabwe Government promoted the growth and consumption of small grains suitable for semi-arid conditions through incentives such as competitive pricing, and including pearl and finger millets on the list of controlled products through the Grain Marketing Board Act (1984). However, years after these small grains had been discouraged by the colonial administration, people in the communal areas now preferred maize, and the cultivation and consumption of small grains was limited to cultural events such as rituals of appeasement and rainmaking ceremonies.

Having observed the inadequacies of the community development approach under colonial rule and the need to include women (who had fought alongside men in the struggle for independence) in the development of the country, a ministry of Community Development and Women's Affairs was set up. The intention was to facilitate development by involving communities through structures such as the village development committees (VIDCOs), the ward development committees (WADCOs) and district development teams comprising membership from government departments dealing with rural and social development, education, health and national affairs.

The post-independence Zimbabwe Government's community development approach did not adequately address the needs of the rural population, based mainly as it was on the group or co-operative approach. Despite the good intentions of the bottom-up approach, which worked through the development structures in the villages and wards up to the district level, the approach did not work, partly due to the modest funds allocated to the Department of Community Development. As was the case in Tanzania, there was effective mobilization for popular participation, but without the funds to plan and finance projects, because the district councils could not raise enough revenue to support the self-help movement.

The other policy decision that was meant to increase participation in commercial production by the previously disadvantaged blacks was the land reform and resettlement programme. In 1981, a Ministry of Lands, Resettlement and Rural Development was established to implement land reform and settle 162 000 families over a period of three years from 1982 to 1985. A number of families were given 'outgrower' plots of about ten hectares each located near an estate owned by the Agricultural and Rural Development Authority (ARDA), and were given support to produce a common crop with the estate. However, this land reform did not benefit a significant number of the communal farmers.

The Zimbabwe Government also tried to improve agricultural production amongst the resource-poor blacks through institutional changes. These changes included measures to scrap separate development and enact legislation that promoted black farmers' participation in the money markets. For example, institutions such as the Agricultural Finance Corporation that formerly provided agricultural services according to race were merged. Credit, extension, research and marketing institutions were restructured to enable them to cater for the communal farmers as well. Perhaps the major achievement of the restructuring programme was the provision of marketing outlets for grain and cotton in communal lands, which increased from three in 1980 to 13 in 1985. During the same period 55 buying points were set up to alleviate transport and access problems that dogged the communal farmers (Rukuni and Eicher 1994).

As part of the restructuring programme, the Department of Research and Specialist Services in the Ministry of Agriculture introduced on-farm research, surveying in communal lands, agro-forestry and research on small livestock. Despite all these efforts by the Zimbabwe Government the communal lands have not changed significantly. Extension and research services are still largely inappropriate as they fail to acknowledge and build on the farmer's knowledge; productive land is still a scarce commodity for the majority of the communal people, and as a result, the late 1990s have resulted in more aggressive methods of communicating to the government the need for land redistribution by land-hungry communal farmers. During the last quarter of 1998 government ministers spent most of their time addressing communal farmers who had turned to squatting on neighbouring white-owned commercial farms.

This discussion of the recent history of African agriculture and the deterioration of conditions for black farmers in the communal lands under colonial and post-independence rule indicates how farmers have lost faith in their existing methods, as well as having little belief in the advice offered to them by outsiders, whether they be agricultural extension officers or NGOs. This was the situation of the farmers in Chivi when they were approached by ITDG's project officer in 1990 to start a project – and their polite scepticism is apparent in their comments about the first meeting (see Box 4.1).

A 'farmer first' approach

When ITDG first began to plan its food security project in Chivi, the approach it adopted was based on lessons learned elsewhere. ITDG's work with pastoralists in Turkana, northern Kenya, demonstrated the importance of participatory development projects, based on an open-ended process of dialogue with the community, or 'putting the farmers first'. This means that the technical aspects of the project were not pre-determined or imposed on the community, the only restriction being that they should contribute towards achieving food security. Two key features of the Turkana work underpinned early project plans: responding to a range of community-identified needs and the importance of working with and strengthening existing local institutions. This led the project to work directly with local institutions such as farmers' clubs as the key partners, rather than through an intermediary organization. These local institutions were to be encouraged to identify their own priority needs in relation to food security, and to plan and implement activities to meet these needs.

Staff of the government agricultural extension service, Agritex, were consulted and involved from the start. In fact, the earliest plans for the project arose from discussions between senior Agritex officials and ITDG's Country Director in Zimbabwe (himself an ex-director of Agritex) about the need to explore alternative strategies for agricultural extension in the communal areas.

ITDG set the following broad objectives for the food security project:

- To increase household food security through improved agricultural production;
- To strengthen local institutions and enable poor farmers (men and women) to articiulate their priorities and control productive resources;
- To influence government agricultural policies to be more responsive to the concerns and circumstances of poor farmers.

The food security debate

In the next chapter the beginnings of the Chivi project will be described, but before this it is worth alluding briefly to what the ITDG project staff understood by 'food security'. The overall aim of the Chivi project is to enhance the food security of marginal farmers in the communal areas of Zimbabwe; the term has been defined in many ways. In general ITDG follows Maxwell's definition of 'removing the fear that there will not be enough to eat', encompassing the concept of long-term access to, not just current availability of, food supply.

Within the food security debate, a number of positions have been taken by the major agencies concerned with addressing the world's increasing food security needs. Although the issues involved in this debate are complex, the various positions can simplistically be said to range from those who advocate widespread adoption of northern (high-input, industrialized) agricultural methods in the south, to those who argue for a new 'double green' revolution. The original green revolution attempted to adapt the environment to cater for newly developed seeds. In contrast, the 'double green' revolution advocates the adaptation (by biotechnology and other means) of the seeds to the environment. ITDG, together with many other NGOs, argues for a food security approach that develops resource-conserving farming techniques, since these are likely to be of more long-term use to poor farmers.

The farmers living in Chivi had their own way of putting this. During an evaluation of the Chivi project in 1996, they came up with their own definition of food security – beating hunger.

3

CHOOSING CHIVI DISTRICT

In 1990, ITDG decided to launch a food security programme as part of its work in the country. The organization felt that ten years after Zimbabwe had gained independence people in many communal lands still faced hunger and food shortages. This is particularly so in the remote and often very arid parts of the country, characterized as natural regions IV and V, where droughts occur frequently.

The project chose an area of poor natural resources, but recognized that even in the driest parts of the country, farmers were skilled in surviving. Therefore the project aimed to work with them to build on their knowledge and harness their strengths to ensure food security at household level.

Where to locate the project

In line with its overall objectives, ITDG sought to identify a community with which to work. As a first step, ITDG commissioned consultants to look at the economic situation in Zimbabwe and help them to identify a district in which most of the major characteristics surrounding communal lands in Zimbabwe today are evident. These are:

- Poor soils, which are unable to sustain any reasonable crop returns without application of fertilizer or manure;
- Inadequate rainfall, with drought in almost three out of every five years;
- Land pressure coupled with high population growth;
- Small landholding sizes with some people landless;
- Poor grazing facilities with some people having no draught power;
- High levels of malnourishment.

The consultants identified Masvingo Province, and Zaka and Chivi Districts within the province (see figure 3.1), as the two areas where all these factors and more were present (ITDG 1991a). All of these are the characteristics of a typical communal land in Zimbabwe. The thinking therefore was that if the project was successfully implemented in such a typical communal land, then it could, in the long term, be replicated in other communal areas.

Out of these two districts, Chivi was selected as the one most suitable to work in. It undoubtedly suffers from food insecurity; folklore has it that once upon a time a

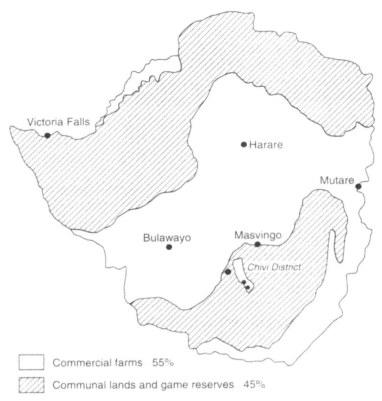

Commercial farms 55%

Communal lands and game reserves 45%

Figure 3.1 Approximate land-use patterns in Zimbabwe

woman from Chivi had nothing for herself and her children to eat, so she boiled some stones and afterwards gave her children the 'soup'. This fable characterizes Chivi as an area where there is general poverty and food insecurity in particular.

Average annual rainfall is 530mm, and drought years (considered to be those in which inadequate rain falls to produce a crop, or < 450mm) occur three years in every five. After the start of the project there were droughts in 1990–91 and 1997–8. The district population is around 170 000 (1990 figures, Murwira 1991a), made up mostly of the Karanga people, a sub-group of the Shona who originally migrated from the north of Zimbabwe. The high population density of up to 70 people per sq km and a population growth rate of around 3 per cent combine to produce enormous land pressure (Murwira 1994a). Average land holdings are around 1.2ha per farmer and are in decline (ITDG 1991a). Before the 1990–91 drought, less than half the population owned cattle (ibid.), and some estimate that up to 90 per cent of cattle died during the drought (Mulvany *et al.* 1995).

Subsistence agriculture forms the mainstay of the household economy for most families in Chivi. However, additional income generating activities play a significant part, such as trading in clothes and food, sale of agricultural surplus, gold panning, crafts and pottery production. Furthermore, remittances from family members working in the urban centres are increasingly vital for rural families' survival.

Having identified Chivi as the district in which work was to be initiated, ITDG carried out a survey to make an inventory of all institutions operating within the district. The purpose of this inventory was to enable ITDG to be familiar with who was working in Chivi, what kind of work they were doing or had done, and ultimately to ascertain how these institutions could collaborate (or might conflict) with ITDG.

Chivi district is one of the biggest in the country. There was no way, even with the best of intentions, that an organization like ITDG could work in the whole district at once. It was necessary once again to narrow the focus to a specific ward. A ward is the second largest unit for community organization in Zimbabwe; it is made up of six villages, the village being the lowest level of community organization. The criteria for selecting the ward were:

- Dry conditions (regions IV and V);
- Remoteness from large business/administrative centres;
- Farmers growing very few cash crops, subsistence farming being the main means of survival;
- Very little or no NGO activity;
- Underdeveloped infrastructure, e.g. roads, schools, etc.

Box 3.1 Ward 21 – the 'dark corner' of Chivi

I think our ward was chosen because we were really the forgotten people. We just used to hear that such and such an organization is at Ward 8 or another ward nearby. But we did not see many of these people. The only ones we ever saw were the ones (Redd Barna) who constructed this community hall. Even government officials never used to come here. There are no good roads here, no nice houses for them to sleep in and no wonderful projects. We were just people [i.e. poor people].

Mrs L. Chiza, farmer, Ward 21

Ward 21 within Chivi District was ultimately selected as the one that fitted all the criteria cited above. The area is also known as the *chomuruvati* area, the 'dark corner'. The selection of Ward 21 was made through discussions with the district administration officials, agricultural extension workers at district level, district councillors, and other key government institutions working in Chivi District.

Chivi Ward 21

In Zimbabwe, 70 per cent of the population resides in rural areas and the majority of rural households are female-headed, partly due to rural–urban migration, a trend which started during colonization when only men were allowed to work and stay in the urban areas. In Ward 21, 16 per cent of the households had female heads. At the time the food security project started, the population of Ward 21 was around 1300 households, with an average family size of approximately seven. When a wealth-ranking exercise was carried out in the two poorest villages in Ward 21 (described in more detail in chapter 4) a large proportion of villagers came into the poorest of the four categories (see Table 3.1).

Table 3.1   Poverty distribution in two villages of Ward 21, taken from villagers' wealth-ranking exercise*

Wealth rank	Description by community members	% of population
1	Enough assets (cattle, ploughs, carts, etc) for own use or to share	21
2	Enough assets for themselves but not to share	22
3	Some assets but no livestock for draught power – which they borrow	21
4	Own very few or no assets: considered 'lazy', 'helpless', or 'stupid'	36

* (Farrelly 1995)

Farming in Ward 21 is largely subsistence agriculture for home consumption. The main source of monetized income in Ward 21 was from vegetable gardens, which earned between Z$120 and Z$1200 (roughly £8–£80 or US$12–$120) per year. Gold panning was the most profitable activity, but this is very dangerous and usually undertaken without registering with the authorities, and hence is illegal. Some families depended on their children employed in towns and cities in various parts of the country for income from remittances. Most households did not have livestock, and cattle were expensive and rare. There were more donkeys than cattle to provide draught power and transport. The food security situation was tenuous and crop failures were very frequent, and so drought relief from the state and other organizations had become an important source of food. Meat was a luxury for most people.

Agriculture in Chivi

In spite of the general poverty at the start of the project in 1991, there is evidence from early settlers that farmers in Chivi were experienced agriculturists in the nineteenth century.

> Travellers who passed through Chivi in the nineteenth century recorded an intensive system of cultivation of valley wetlands (dambos), combined with some limited cultivation of upland areas and red soils when conditions were right. Mauch, travelling in 1871, noted: 'Between the kopjes (rocky outcrops) or on other sides of a small perennial river are the furrows of cultivated ground that is intended for growing rice.' Similar impressions were documented by Thomas Leask in 1867: 'The hill was surrounded by rice gardens. These rice fields are in low swamp places and, the better to hold water, they are under ridges'(quoted by Wilson, 1990). W. H. Brown, travelling with the pioneer column in 1890, reported: 'A thickly populated region and the valleys filled with fields of rice, meallies (maize) and kaffir corn (sorghum). The people ... built their houses in rocky fastnesses' (quoted by Wilson 1990).
>
> (Scoones and Hakutangwi 1996a)

Wilson argues that hoe cultivation of rice and root crops in valley wetlands was the most important farming method and labour was mobilized through lineage clusters (ibid.). However, from the 1940s, a series of land-use planning initiatives and enforced natural resources legislation increasingly constrained farming opportunities. The Chief Agriculturist of the Native Department, Emory Alvord, in 1929 initiated the centralization policy in nearby Shurugwi. 'The policy aimed to modernize local farming systems, which were perceived to be backward, unproductive and unsustainable. Farmers were encouraged to adopt a package of technical advice based on improved varieties of grain and cash crops, rotational cropping and the manuring of arable lands'.

In 1902, there were only 18 ploughs in Chivi District, 1300 a decade later and over 5000 by 1930. The combined effects of the shift from wetlands to dryland cultivation and the movement of Africans into the 'reserves' and the introduction of technologies such as ox-drawn scotch carts led to accelerated land degradation in Chivi. For example, in 1926, in the whole district, the native commissioner recorded three scotch carts, by 1939 there were 54, nine four-wheel wagons, four motor cars and two lorries (ibid.).

The 1950s saw the beginning of the adoption of other new technologies for boosting production including hybrid maize, a variety of cash crops such as cotton, and inorganic fertilizer. Although maize was vulnerable to drought conditions, the major yield breakthrough achieved with the introduction of SR52 and its successors led to the erosion of the drought-resistant varieties of flint maize and small grains such as pearl and finger millets. 'By 1970 nearly everyone was planting at least some hybrid maize in Chivi' (ibid.).

Agricultural extension services have a long history in Chivi. The first demonstrators arrived in Chivi in 1927 when two were appointed by Alvord to work alongside 10 demonstrator farmers. A third demonstrator was posted to the district in 1939. The master farmer training scheme was introduced in the district around this time. The master farmer scheme and much of the extension support focused on relatively resource-rich farmers who were able to make use of information offered and adapt the technologies recommended. Today the ratio of extension workers to households is 1:800, and the uptake of their recommendations by communal farmers is poor.

The master farmer scheme continued after independence and in fact when ITDG moved into Chivi one of the institutions they decided to work with was the master farmers' clubs (see chapter 4).

Institutions in Chivi – traditional and modern

The institutions operating in Chivi, and the people who wielded influence within these organizations, were important considerations in the design of the project. The institutional framework in Chivi involves a combination of formal and informal organizations. The district is divided into wards, which are made up of a number of administrative 'villages'. These are administered by the WADCO (Ward Development Committee) and VIDCO (Village Development Committee) respectively, the chairs of the VIDCOs forming the WADCO, together with the elected ward councillor. The district council is made up of all the councillors and the district administrator, the senior civil servant in the district.

The village leadership consists of *sabhukus* (kraalheads), headmen and chiefs. The system of *sabhukus* was in fact a colonial invention, based on the lineage system that preceded it. Traditional lineage leaders have historically been responsible for land allocation, and every household in the area was related in some way to them. Lineage forms the core of traditional Karanga society, which has meant that immigrant families are often discriminated against in the distribution of land and other issues. In contrast to the traditional system, the VIDCOs and WADCOs, established following independence, are commonly made up of younger men from immigrant families. Since the VIDCOs were given powers to allocate land, there is constant rivalry and sometimes conflict between them and the traditional lineage leaders, in particular the *sabhukus* (ITDG 1991a, 1995a).

In addition to these older leadership forms, there are a number of semiformal or affiliated groups, such as the farmers' clubs (linked to the Zimbabwe Farmers' Union), women's garden groups, church groups, local branch groups of ZINATHA (Zimbabwe National Traditional Healers' Association) and so on.

After consultation with the community, ITDG chose to work with the farmers' clubs and the women's garden groups because they were local institutions that were directly involved with food production (they are described in more detail in chapter 4). The process of making acquaintance with the communities of Ward 21, getting to know their farmers' clubs and garden groups, and assessing their needs is the subject of the next chapter.

4

INVESTIGATING NEEDS

Entering the community

In 1991, ITDG was introduced to Ward 21 by the district Agritex leadership and local government officers. A meeting of all community leaders and community members was called, and about 700 people attended this meeting. Mr Kuda Murwira, who represented ITDG, commented on the warmth of his welcome, but also noted that the villagers were perplexed about what ITDG had to offer (see Box 4.1). They had met other NGOs before, but were used to being presented with a pre-formulated programme, with many concrete benefits. Instead, ITDG's approach was to draw out the people's needs, and help them to develop their own resources for meeting them.

Some members of the community were also suspicious, thinking that ITDG was going to collect information about them and then use it against them; they thought ITDG wanted to 'spy' on them. There were a number of very outspoken people at that first meeting who seemed to have the final word on every issue, and ITDG staff decided that rather than have a long discussion with them in front of everyone, it was perhaps better to hear what their reservations were individually and try and overcome their resistance.

A week was spent visiting some of these outspoken people and listening to their reservations. This was important because ITDG preferred not to confront them in a public meeting, but instead wanted to talk to them more privately. Giving people individual attention was also seen to be a better way of making them feel comfortable enough to talk. It was also felt that if the project was going to work in this ward, everybody in the community had to participate: having some disgruntled people within the community, few as they might have been, would have actually resulted in the disgruntlement surfacing later and probably disrupting the work.

It was during the process of talking to the few community members individually that ITDG learned that most of the people were already cynical about 'outsiders' who had come into their community, told them what they wanted to do, did whatever they wanted and left. The community felt used by such organizations, and at the end of it they did not see any tangible results. ITDG took this

Box 4.1   Reactions to the first meeting

When you come into the home of someone you do not know for the first time, you have to make sure that your entry is well planned and you will be welcomed. You cannot just walk straight into my yard, walk everywhere including over my sacred places, step on my chickens, and then tell me that you have come to help me solve my problems. You have to come in nicely.

Mr S. Masara, farmer, Ward 21

The meeting was very good and the reception we got was also very warm. Zimbabweans are very polite people so in most instances they will welcome you very nicely, even if they are not happy with your presence. The greatest difficulty was in explaining what we wanted to do with the community. We were not like other NGOs telling them the objectives, the activities that were going to be undertaken and what we expected from the community. We were not offering any money either. Some of the community leaders expressed their unhappiness. They told us that they had aired their problems many times before to various agencies and nothing had come out of it.

Kuda Murwira, ITDG project officer

This man, Murwira, he said I have come to work with you to end hunger, but I have no money to give you, no free food, nothing. We are going to work together to think about what our problems are and how we can solve them. Then I thought, this man must be mad, or he thinks we are stupid. Why is he wasting our time like this? It was only later that we realized his approach was good. Because if he had promised us many things and failed to deliver people would have been very disappointed.

Mr S. Masara

opportunity to explain to the people that this is precisely why they now sought to work with the community in a different manner: identifying problems with them, devising solutions with them and simply helping them to help themselves.

This approach of talking to some of the 'spokespersons' on a one-to-one basis helped. At the next meeting, it was these spokespersons of the community who now became spokespersons for ITDG. The only nagging issue was that ITDG was not offering any financial or material support.

At this second meeting the main objectives were to seek the community's agreement to work with ITDG and the selection of two villages in which the project would first be implemented. Villages C and E were chosen for reasons listed below; letter names were given when new villages were established to facilitate development in 1984. These new villages were largely imposed on the people, and the people chose not to 'own' the new structures by giving them local names, so they are still known by their letters.

- According to the community leaders, these villages had a higher concentration of resource-poor people than the other villages in the ward;
- They fitted the general criteria used in identifying the ward;
- Village E in particular, although resource poor, was very well organized; there was an active village community worker (VCW) who managed to mobilize people for all kinds of activities. The village was also quite small, with fewer than 90 households, making it very cohesive. For this reason it would therefore make the work of ITDG easier since people could easily mobilize themselves to work together;
- In contrast, village F was very large, with 280 households; because it was a newly settled area, people did not know one another very well and the village community worker had recently died. As ITDG wanted to work through local institutions and strengthen these for sustainability of the project in the long run, it was thought that village F would not be ideal for a start.

IT staff discussed these criteria with the community leadership, and together it was therefore decided that villages C and E would be the pilot villages.

Box 4.2   Reactions to the second meeting

The people chose villages C and E as the two to start off with. I think it was because some of the people thought, ah, since this organization is not giving money or 'things', let us kick them to those ones over there who have absolutely nothing.

Mrs L. Chiza, farmer, Village E

We talked about all the leadership in this ward, the small groups we have here and how we work, our problems in these groups and what was good about them. We have many groups and many leaders in our areas. Not all are good; not all are bad. It was very good to do this because we could then say, this is what is good about this and this is what is bad. It was not about criticising anybody at all. We wanted to build each other.

Mrs Anna Gatawa, farmer, Ward 21

Identifying partner institutions

As one of the major objectives of the Chivi project was to strengthen local institutions, the next part of the process was to identify those institutions that ITDG could work with in planning and implementing the project. In some of its work in other countries ITDG had learned that creating new institutions around a specific project could lead to problems once the project was completed. When this happened the new institutions often simply died when the external facilitating agency withdrew. ITDG also realized that it was important to assess the strengths and weaknesses of existing institutions so that decisions could be made by the community themselves as to which would be the best institutions to work with and ensure that the momentum would be sustainable even after the withdrawal of the facilitating agency.

In almost every community in Zimbabwe there are institutions of various kinds. Through these people organize themselves to tackle certain problems, or simply to socialize with one another. Ward 21 was no exception.

An institutional survey was carried out at ward level to do four things: first, to identify institutions (both formal and informal), that existed and what role they played in the community; second, to assess the strengths and weaknesses of each one with a view to strengthening some of these if need be; third, to discover resources that existed within the community which could be harnessed for development; and fourth to identify the relationships between these institutions in order to avoid deepening any conflicts and to strengthen already strong relationships.

The institutional survey was carried out by the ITDG project officer, assisted by the VCWs (who are paid a monthly stipend for community mobilization activities by the Ministry of Co-operatives and Community Development). Two men and four women were interviewed from each of the six villages in the ward, as well as local civil servants and community leaders, over a period of six weeks. The leaders of the institutions were interviewed at these very early stages to help them to reflect on their own institutions. Ordinary members of the clubs were selected in order to bring out issues as seen from the ordinary members' perspectives, especially issues of leadership. Non-members of these clubs were also included in the survey: they were asked why they were not members, and what were their objective views of the institutions, as outsiders.

The institutions in Ward 21 were quite varied and included the following: traditional leaders (chiefs, headmen and so on), churches, VCWs, ward community co-ordinators, extension workers, traditional leaders, farmers' clubs, garden groups, VIDCOs and WADCOs.

In order to make the final selection of local institutions to work with, the objectives of the food security project were revisited. The following criteria were seen as important in deciding which were the best institutions to work with:

- Their activities related to food production;
- Absence of conflict between their activities and traditional practices;
- A truly democratic and representative leadership;

- The active participation of women, especially in decision making;
- A membership including otherwise marginalized people;
- No bias in favour of one ethnic group;
- Evidence of sensitivity to indigenous knowledge and practices.

The garden groups and farmers' clubs were identified by the community as the most promising institutions dealing with food security within the community, indeed perhaps the only functional institutions in Ward 21 outside the political and traditional leadership structures. ITDG project staff made this initial selection, which was then confirmed by the community itself at a feedback meeting held to discuss the findings from the local institutional survey. It is possible that the community selected these organizations because they were familiar with them, and that they had survived with minimal outside intervention. One of ITDG's objectives was to strengthen people's own institutions so that they could implement whatever project the community agreed upon and also sustain the project even after ITDG had left. It was also felt that these institutions were neutral in the power struggle over land allocation which affects the *sabhukus* and the VIDCOs. Selection of either of the latter organizations as key partners might have alienated the other. Other institutions, such as the VCWs, were there to support the activities of these groups.

One of the reasons for the selection of farmers' clubs was the potential link with a national organization, the Zimbabwe Farmers' Union (ZFU) whose membership consisted of smallholder farmers in rural areas throughout the country. ZFU catered for the black majority, while the Commercial Farmers' Union (CFU) dealt mainly with the white commercial farmers.

The composition of the farmers' clubs posed a problem, however. In an effort to improve agriculture among the rural farmers, the ZFU had been instrumental in the formation of farmers' clubs, based originally around at least two master farmers. Master farmers undergo a two-year training course in approved farming techniques for which they are given certificates by Agritex, and the farmers' clubs tended to be drawn from an élite of wealthier farmers from a number of villages. With their present membership it was difficult to see how the farmers' clubs would directly benefit poorer farmers. However, at the time of the start of the project, ZFU changed their policy and encouraged groups to be formed around the *sabhukus*; ZFU also planned to remove the requirement of two master farmers per group. This, the ZFU hoped, would increase membership (there were only 6000 individual members out of about two million communal area farmers). Community leaders and existing club members welcomed the change as a positive way of facilitating greater participation by the whole community, especially the more marginal members. The change was also viewed as a way of forging a constructive link with traditional leaders.

The garden groups, whose membership was 90 per cent female, were selected partly to facilitate women's participation in the project. The garden groups had been set up under the then Ministry of Community Development and Women's Affairs whose Community Development Fund provided grants for the groups. However, the grants were often either misdirected or misused and consequently some of the groups disbanded. Others continued to work together to produce much-needed vegetables for household consumption and sale. Vegetable gardening contributed significantly to household food security and provided some income but was largely overlooked by the extension services, so it was considered that selection of the vegetable garden groups would contribute to raising the status of the activity.

It was clear that both clubs needed ways of making them less status-oriented and more functional and helpful in addressing the food security problem.

Not overlooking the marginalized

During the first few weeks of the project, the project officer spent days visiting households, talking to the farmers and sometimes joining in whatever activity they were engaged in. It was decided that for the ITDG project officer to be accepted, and for the community to feel they owned the project, it was necessary to put aside the urgent need to start implementation and concentrate on the process of entering the community, getting rid of the people's suspicions and building up their confidence, particularly the confidence of the marginalized. Getting to know the community in this manner laid the basis for a good working relationship between project staff and the community.

Box 4.3   Including the poorest

NGOs often say that they are working with the poor and the marginalized. Yet they turn up in a village and think that all rural people are the same. But within many communities there are differences in wealth, status, and even perceptions of one another and their problems. It is important to look very closely at a community and see the differences among people to ensure that the poor are not further marginalized.

Kuda Murwira, ITDG project officer

In the past Agritex workers concentrated on the richer farmers. They had the money to buy implements and to try out new technologies, and they were literate. In addition an extension worker would want to impress his superiors when they came to visit his area. So what better way to do it than to take them to a nice house, and a big field with thriving maize etc.

Mr Butaumocho, Agritex extension officer

Box 4.3 Including the poorest (continued)

This was all new to me. In the past there were farmers who were really farmers. These are the people who mattered. Some of us were forgotten over there near the waste-bin. It was through this project that we began to be seen as people who could think and say something useful. When some people saw a poor person they would say here comes that one with her problems. They thought you just wanted to borrow or ask for things. But they saw that we can think and we can be involved together with them in developing ourselves.

Mrs L. Chiza, farmer, Ward 21

The project officer also observed what had taken place in the community in terms of development assistance from the government or other NGOs. Sometimes when he visited homesteads of the poorer members of the community he found that they felt left out and believed that development agents and development were concerns of government and the well-to-do members of the community, including the master farmers. Extension workers were accused by them of choosing to work with this rural élite.

In order to strengthen the weaker and the poorest of the community and minimize the danger of the project being opposed by the powerful, the more prosperous farmers were involved in the studies leading to the design, planning and implementation of the project. It was recognized that in any development project, people participate because of how they think they will benefit from being involved. The poor may participate because they hope to achieve food self-sufficiency with some surplus for sale, whereas the rural élite hope to gain popularity, respect and acceptance by the poorest of the community. In Ward 21, the élite needed recognition by the rest of the community, while the majority needed enough food for their families. They had nothing to lose by participating.

As they gained confidence in themselves, the poor realized that each person had certain competencies that could be beneficial to the individual and to the community – no individual was completely useless. They also recognized their own power to bring about change in their lives. The élite needed to maintain their leading positions in all aspects of life – in business, politics and even in socio-cultural situations such as custodianship of rituals and ceremonies, and the poor realized that they could withdraw their support for any idea or from an individual – in other words, that the élite needed them.

At the outset, and before the project could be over-influenced by richer people, it was important to identify the poorest people to find out how they defined their problems. In villages C and E of Ward 21 there are clear differences between the people; for example, some have big houses, while some have one hut. Some have oxen spans for ploughing, while others do not have a single chicken.

Wealth ranking

In order to assess the needs of the communities in the two villages it was essential for ITDG to:

- Identify the real needs of resource-poor people;
- Identify what the community itself saw as poverty and its manifestations;
- Note what perceptions the community had of resource-poor people and how resource-poor people saw themselves;
- Identify the variables that could make a family resource poor, such as having no adult male in the household.

In carrying out this part of the work ITDG used a number of methodologies, the main one being a wealth-ranking exercise. One of the benefits of this exercise was that, once ranked into wealth groups, a stratified sample could be drawn for the needs-assessment exercise which followed. In order to facilitate wealth ranking, all the households in the two villages were listed.

After this a community meeting was called to explain this part of the process and its objective. A number of volunteers were asked to help the ITDG team with the wealth-ranking exercise; they had to be people who knew every individual and every household in the community. Two men and two women were eventually chosen; a community development worker who works in both villages was also asked to do the ranking for both villages as she knew all the households listed fairly well. It should be noted that it was difficult to get people who knew every member of the villages very well.

Each of these volunteers was asked to sort the names of the community members in four different piles according to how they 'ranked' them, and also to explain why they had ranked them in that way. In addition, the whole group of volunteers discussed all the piles to try and see if there were any major differences in how each person had ranked the households. The four main wealth ranks identified were as follows.

- *Wealth rank 1*   This category comprised people who have: enough cattle, ploughs and other agricultural implements, enough food for themselves and also to sell sometimes, a good homestead (with at least a spare room for visitors), and so on. This group was described by most people as 'master farmers'. This group was also defined as having enough assets for themselves and also to share with others.
- *Wealth rank 2*   The best description for this group was that they have enough assets for themselves but not enough to share or lend out to others.
- *Wealth rank 3*   People in this wealth rank were described as having some assets but not many, for example, they do not have draught power, but they are able to borrow and can survive and thrive.

Photo 4.1   Community members sort households into four wealth ranks
ITDG/Kuda Murwira

- *Wealth rank 4*   This group was seen as being at the bottom of the wealth ranks. Many of the people in this category were those with no assets or very few resources.

The wealth-ranking exercise (the results of which are shown in Table 3.1 on page 24) was revealing: it showed the values people at community level place on certain resources and assets and also how they viewed one another. For example, having a house with an asbestos or iron roof was seen as a sign of high status. Men tended to value or place emphasis on resources like draught power and farming implements, while women valued a good homestead with an extra bedroom for visitors, and so on. The ability to send one's children to school was also a value ranked highly by the community members.

Perception of others was another significant insight from this exercise. For example, people in wealth rank 4 were described as 'lazy, stupid, helpless or shaky' by other members of the community. These observations were useful in informing ITDG about some of the opinions and misconceptions about poverty and poor people which existed in the community, which could be addressed in the implementation of the project. This information would also be useful in the evaluation and monitoring process, as ITDG and the community sought to

Box 4.4   'Ask me what my problem is'

If you came here and said Mai Chiza you are poor, you need a dress, and you gave me one, that may not be what I want. I may take that dress and smile very nicely and thank you very much, but after you go I will turn around and say 'that idiot'. I may take the dress and just hang it up on a stump. Or I may wear it, but I will wear it back to front. So you have to ask me what my problem is and what I want to do about it. Then we agree on how you can help me if you want to and if I would like you to.

Mai Chiza, farmer, Ward 21

There is a well-known joke about a development worker who went to a village to undertake a survey. He grouped women together and asked them what work they did. They told him how they took care of their children and husbands, fetched water and firewood, weeded the fields and did numerous other 'domestic' chores. They concluded by saying that they did not really work. Upon which the development worker said, 'Good, I was just checking to make sure' and ignored them thereafter.

assess the impact of the project, and find out who had participated and whether perceptions had changed.

Needs-assessment survey

Before any actual work could begin in Ward 21, ITDG had to work with the community to identify what exactly their needs were and how these could be addressed. It was essential for ITDG together with the community to define what exactly the problems in the community were, in what form they presented themselves, why the problems were occurring and ultimately how they could be solved (see Mai Chiza's words in Box 4.4).

Building on the results of the wealth-ranking exercise, the next part of the process was intensive interviews with individual families from the different wealth ranks in a community needs-assessment survey. The wealth-ranking study ensured that for this part of the process, community members from all the different wealth ranks would be involved – with special emphasis on the resource poor.

A representative sample of 10 per cent (30 households out of a total of 416 in the two villages) of the total number of households was chosen for the intensive door-to-door survey. These households were in all the wealth ranks, with more bias on the poorer households (70 per cent of respondents were chosen from wealth ranks 3 and 4). Particular attention was also paid to female-headed

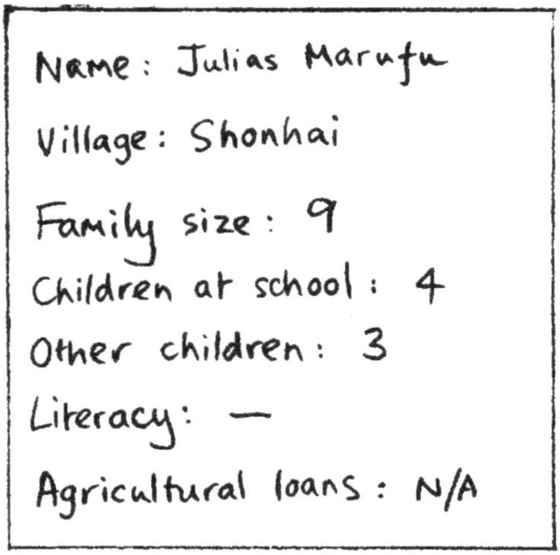

```
Name: Julias Marufu

Village: Shonhai

Family size: 9
Children at school: 4
Other children: 3
Literacy: —
Agricultural loans: N/A
```

Figure 4.1   Example of a wealth-ranking form

households and to capturing the voices of women within the households. Again this was based on the recognition that very often men tend to be seen as heads of households and therefore spokespersons of the family. ITDG had therefore to ensure that women, who in the case of Zimbabwe comprise the majority of residents in communal areas and who do the bulk of the work in the area of food security, were involved at all stages of the project.

The door-to-door survey revealed not only the needs of the people but the following issues too.

- It confirmed the wealth rankings that had been done by a few community members. In a few cases, however, the wealth rankings were found to be inaccurate. For example, some widows had simply been graded to the fourth wealth rank because they were widows, yet they had more assets and were thriving better than had been assumed.
- The food security situation was very tenuous in the area and drought and crop failures were very frequent. Traditional crops like sorghum had been overtaken by maize even though the crop did not do very well in the climatic conditions. Drought relief from the state and other organizations had become an important source of food.
- Meat was a luxury for most people.
- Sales from vegetable gardens were the most common sources of income for poor people. However, this did not pay very well, with average earnings ranging between Z$10 and Z$100 a month. Many people did not have surplus income even for basic necessities like clothes or sending children to secondary school, since most of the little earned went towards food purchases.

- Some families got income from their children who were employed outside the village.
- Women tended to do a lot more of the intensive weeding than men. Similarly gardening which brought in income and extra food or protein was done by women.
- Most households did not have livestock. Cattle were now expensive and rare and donkeys were more common.
- Farming implements were highly valued and families tried to acquire some; even if, for example, they did not have cattle they would have a plough. Hoes, harrows, scotch carts and other farming implements were priorities for most people. The poorest of the poor had a few hoes at least and many depended on borrowing the bigger implements from neighbours or extended family members.
- Many of the poorer people did not belong to the farmers' clubs as the requirements and the expectations were too high. For example, one had to pay a membership fee, and to invite the club to come and help you on the field you had to provide a goat/meat and all the food.
- Most of the poor people had never worked with locally based agricultural extension workers as they tended to work with wealthier farmers (see Box 4.3).

As well as the household survey, group discussions were also held with two garden groups. Participatory rural appraisal (PRA) techniques such as seasonal calendars and transect walks (see figure 4.2 and 4.3) were used to obtain information on workload, income and expenditure and food supply.

Discussions revealed how farmers perceived the state of their land, and in particular the decline in their resources over the past 20 years; these are

Table 4.1   Farmers' perceptions of the state of their resources

Resource	Present state compared with 20 years ago
Soil	Exhausted, fields eroded, lots of gullies have formed
Water	Most wells have dried up
Vegetation – trees	Only a few left, no more vegetation cover on exposed hills.
– other	Only thin cover left
Livestock	Only a few left (due to 1991–2 drought)
Wildlife	None left
Population	Considerably increased

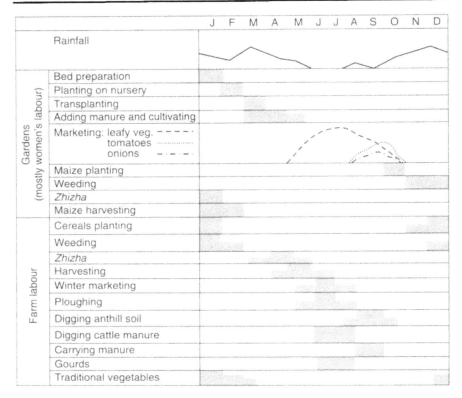

		J	F	M	A	M	J	J	A	S	O	N	D
	Rainfall												
Gardens (mostly women's labour)	Bed preparation												
	Planting on nursery												
	Transplanting												
	Adding manure and cultivating												
	Marketing: leafy veg. − − − − · tomatoes ············ onions − · − ·												
	Maize planting												
	Weeding												
	Zhizha												
	Maize harvesting												
Farm labour	Cereals planting												
	Weeding												
	Zhizha												
	Harvesting												
	Winter marketing												
	Ploughing												
	Digging anthill soil												
	Digging cattle manure												
	Carrying manure												
	Gourds												
	Traditional vegetables												

Figure 4.2   Seasonal calendars demonstrate what time of year carries the heaviest workload, for women as well as men

summarized in Table 4.1 (Hagmann and Murwira 1996). Most people realized a need for resource management. Their comments expressed helplessness, however, and that they had not been able to organize themselves to define and implement appropriate laws to counter the decline. One comment was: 'there is nowhere we can move to after the final degradation of our land'.

Throughout these discussions the most important point the community members emphasized was that they wanted to have enough food to eat with their families and be secure, with a little extra from one year to the next. The lack of water or rain was identified as a key obstacle to the attainment of this dream: water for gardening, farming and for the animals, and for the people themselves.

The needs-assessment survey not only identified some of the key problems that the community was concerned about but it also brought out information on how people live and survive. It also brought out people's value systems and their priorities in life.

	Section A	Section B	Section C

	Runde River along Tende–Chasiya road	Mr Haruzivi's farm	Musvovi school to edge of Munaka
Soil	• loamy soils • range of mountains giving rise to wetlands • gentle slope (2–7%)	• sandy soils derived from granitic rocks • amarula tree as indicator of very poor soils • fairly flat terrain • slope (2–3%) • indigenous soil and moisture conservation techniques	• sandy loam soils • poor soils • fairly gentle slope (2–5%) • visible erosion
Water	• Runde River main source of water for livestock • reliable borehole • unprotected deep wells, some with permanent water supply	• Save River main source of water for livestock • reliable borehole (community) • several unprotected deep wells	• river main source of water for livestock • borehole for community • several unprotected deep wells
Vegetation	• riverine forest along the Runde River • mountainous range characterized by brachstigia spp. ('mubondo, msasa) • grasses (panicum and eragrostis dominant)	• mainly arable land • scattered trees characterized by 'mupfura and 'mopane species • grasses (panicum is the dominant species)	• vegetation is typically tree bush savannah type • combretum dominant tree species • grasses (panicum dominant)
Socio-economic indicators	• grass-thatched huts with few asbestos-roofed houses • fowl runs for turkeys and chickens	• grass-thatched huts	• grass-thatched huts with some asbestos • Border Munaka School

Figure 4.3 Village transect walks can reveal changes in the natural resources landscape, as well as the problems and opportunities they present (continued overleaf)

Achievements over the past 5 years	• borehole sunk at new school at slab level • marketing of cotton as a cash crop • dam site planned by Agritex	• achievment of food security • marketing of cotton and groundnuts as cash crops • establishment of consolidated gardens and sale of vegetables	• a few asbestos houses built • attainment of food security • marketing of produce
Forestry and agro-forestry checklist	• no plantations • home gardens • homestead gardens with mango, paw-paw and guava trees	• no planatations • consolidated gardens • agro-forestry, mostly indigenous trees	• scattered eucalyptus trees • home gardens • fruit trees at homesteads and schools
Problems	• shortage of draught power • lack of adequate 'protected' water sources • transport (nearest bus stop about 10km) • marketing of produce to grain marketing board (GMB)	• shortage of draught power • financial problems (to buy adequate fertilizers)	• shortage of draught power • land shortage resulting in streambank cultivation
Opportunities	• beer brewing • soil and moisture conservation techniques • marketing of cash crops	• vegetable growing and marketing • soil and moisture conservation techniques	• marketing of produce to GMB • marketing of broilers and small stock

Figure 4.3   (continued)

5

PLANNING PROJECT ACTIVITIES

By the end of 1991 the ITDG project officer had gained a lot of information about the communities of villages C and E in Ward 21 through the wealth-ranking exercise and the needs-assessment and institutions surveys, and they in turn had picked up a lot about his approach. The main aims of the next meeting were to come to a consensus on what the people identified as their main concerns, and to help them to plan to meet these needs.

Listing priorities

A community meeting was held to feed back the results of the needs-assessment survey. Representatives of the various institutions who were working at the local, village level were invited, including VCWs, *sabhukus*, the Agritex extension worker, farmers, master farmers, garden group members and community-based family planning distributors. The aims of this meeting were:

- To feed back to the rest of the community the issues and needs identified;
- To enable the community to prioritize its needs in the area of food security;
- To clarify with the whole community the underlying causes of the problems identified;
- To link the problems identified with the best possible institution(s) to address them;
- To draw up a schedule for the work to be done in addressing the identified needs;
- To reach consensus on ways of working with and integrating the poor in all activities.

Collective decision making and ownership of the project was an essential element in the Chivi project. It was through workshops such as this one that this collective process could be facilitated. The community meetings also fostered a sense of co-operation among the community members.

Box 5.1   The need for co-operation

We are all individuals with different needs and wants. But in this area one of the things we wanted ITDG to help us with was working together. Before ITDG came each one was doing his or her own thing, but there was no development because as we say in Shona *nzara imwe haitswanyi inda* (one nail cannot squash lice). So in whatever we did we had to agree as a community so that we could develop together.

Mrs Gwatiziva, farmer, Ward 21

The first activity at this workshop was the expression of expectations for the workshop by the participants. Some of the expectations were: to find ways of improving quality of life, to find ways of ending hunger, to find ways of increasing water supplies to the vegetable gardens and to find ways of developing ourselves. From this exercise it was clear that the community was concerned about finding ways of helping themselves overcome their own problems. This was in line with ITDG's objectives which the ITDG facilitators presented to the group.

ITDG presented its objectives as an organization, its history and other information about itself to the participants. Although the project officer had been working in the area for about six months, there had been no formal introduction of the organization, as the emphasis had been on getting learning from the people, doing away with suspicions, building a relationship based on trust and getting the community to accept the presence of the project officer. Now this trust had been gained it was important to tell the people more about who ITDG was and where it was coming from. Since ITDG was seeking a relationship with the community, it was important that both sides got to know each other.

After the introduction of ITDG, participants were divided into their institutional groups, that is, the garden groups, the *sabhukus*, and so on. Each group was to discuss food security needs and to prioritize them; a representative of each of the groups would present their findings during the plenary session.

Many of the groups were composed of a single sex, for example, the garden groups were all women. By encouraging the voicing of views from such groups it was possible for the priorities of women to be recognized, and acted upon, as well as men. This is important since most of the vegetable gardening was done by women and nearly all the dryland farming was carried out by men. So women's main priority was water for domestic use and vegetable gardens, whereas men prioritized water for dryland farming. Failure to recognize the differing issues for men and for women would have resulted in wrong diagnosis and therefore wrong prescription (see Box 5.2). The approach used for prioritization eventually

> **Box 5.2** Taking into account the different roles of men and women
>
> When we went to Ward 21, we did not take with us strategies to develop the area, we wanted to learn with the community and to discover together what would work for Ward 21. So when we realized that there were differences between men and women, and that these were based on use, we decided to work within those parameters so that both men and women could achieve their goals. Had we not done so, we might have recommended the construction of a dam or an irrigation system which would not have benefited the women's source of livelihood, the gardens, for example. Location of such a scheme could have brought about other problems in terms of use, ownership and accessibility by the whole community. An irrigation scheme would not have been possible in the mid-term for various reasons. Besides, we had not budgeted funds for that kind of project.
>
> *Kuda Murwira, ITDG project officer*

resulted in the choice of low-cost water-conservation methods which the community could rely on. Each family could implement and sustain these methods at no cost.

In plenary each group presented its list, after which the whole group discussed them all and came up with a consensus. The priority list from the whole group was as follows:

- insufficient rain for crop production
- the absence of draught power
- not knowing suitable crop varieties
- gardens unprotected from livestock
- pests and diseases in vegetable gardens
- a shortage of land for farming.

The community had never gone through such an exercise – particularly the poorer members. Their involvement and participation in the processes of identifying, prioritizing and finding solutions to the community's problems enhanced their confidence and self-worth. Participants also enjoyed the teaching methods (see Box 5.3). Using techniques like problem trees and web charts to show how one thing leads to another, the participants were able to look beyond the immediate presentation of a problem to its root causes.

In subsequent workshops, participants discussed each problem and identified possible solutions. For each solution they had to discuss what constraints they were likely to face. For example, when water was identified as a priority, dams or irrigation schemes were considered expensive and discarded because

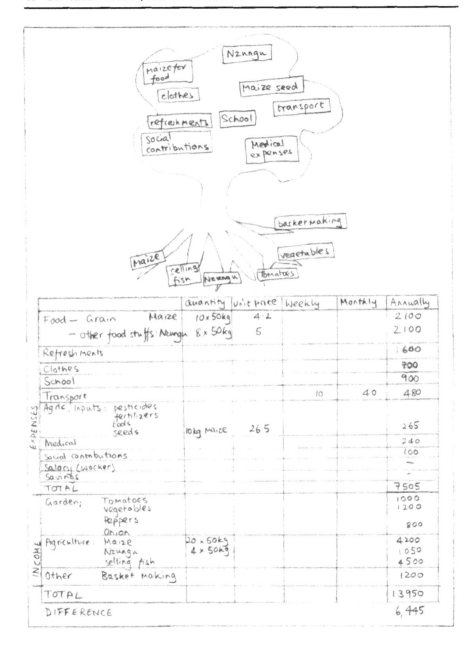

	Quantity	Unit price	Weekly	Monthly	Annually
Food — Grain Maize	10 x 50kg	4·2			2100
— other food stuffs: Nzungu	8 x 50kg	5			2100
Refreshments					600
Clothes					700
School					900
Transport			10	40	480
Agric. Inputs: pesticides fertilizers tools seeds	10kg maize	265			265
Medical					240
Social contributions					100
Salary (worker)					—
Savings					—
TOTAL					7505
Garden: Tomatoes vegetables					1000
					1200
Peppers					800
Onion					
Agriculture: Maize	20 x 50kg				4200
Nzungu	4 x 50kg				1050
selling fish					4500
Other Basket making					1200
TOTAL					13950
DIFFERENCE					6,445

Figure 5.1 A problem tree helps villagers identify their main sources of income and expenditure

the community felt they could not afford such schemes. In the case of contour ridging, the solution was discarded because some farmers did not have the

Box 5.3   Getting to the root of the problem

All of us were and are concerned about the hunger we experience in this area. We may have had different experiences, for example, some people complained about having no cattle, while others were concerned about pests attacking the vegetables. However, it all had to do with food and hunger.

Mrs L. Chiza, farmer, Ward 21

We all knew our own problems. But what we had not discussed in detail were the causes of the problems. So we went through these exercises where we looked at a problem like a tree. On the surface you can see the branches and twigs of a tree and chop them thinking this is the real tree. But no, you must go to the root to see where the tree starts. That is what you must understand. So the next step was to say what do we do about these problems?

We also wanted to say, but if we say this is what can be done about it, who will do it? It was ourselves who were going to do it. So we needed to say can we do it or what are the problems that we will face in those solutions to our problems. We were not just dreaming like children. We wanted to say this is what can be done.

Mr K. Mavhuna, farmer, Ward 21

necessary tools for that kind of ridging. The traditional *humwe* (work party) was revived and modified in terms of the provision of food for the participants. Previously, the poorer members of the community could not organize work parties as they could not afford to feed all the participants.

Prioritizing the priorities

At the second workshop, which took place at the end of 1991, the community leaders agreed to tackle the priority needs in phases, avoiding an attempt to address everything at once and achieve very little in the end.

The first phase of implementation was to address the following problems:

- water for fields and gardens
- pest and disease control, especially in vegetable gardens
- fencing material for vegetable gardens.

On the question of which institution was best able to deal with the identified problems, participants confirmed that farmers' groups and garden groups were the best to address food security issues.

As part of the discussion at these workshops attitudes and practices that marginalize the poor were discussed. These included the opinions that the poor are helpless, they have no ideas, they have nothing (materially), and they do not

Photo 5.1   Planning workshop participants in small group discussions
ITDG/Cathy Watson

participate in community activities (see Box 4.3). In discussion the participants admitted that these perceptions of the poor are what contributed to their lack of participation and marginalization. It was agreed that each institution represented would make efforts to draw in very poor people and discuss with them how they could participate in all activities.

Planning the process

The project officer, Kuda Murwira, had been visiting the villages of Ward 21 for six months by the time these community meetings took place. During this time, as well as his gaining an understanding of the people – their needs, skills, workloads and their institutions – they had got to know him, and begun to accept him. The initial period of learning from the community was essential for the success of this planning process. If pre-formulated projects had been imposed on the community, then they would probably not have been suited to the people or accepted by them, and if the project officer had no knowledge of the whole community he might have been over-influenced by the plans of the few without understanding how these would affect the many.

The role of the project officer was to facilitate the people's own plans, rather than to set his own before them. This did not mean sitting back and allowing them to proceed without any contribution from him, however; in order to draw out the members of the community who are usually quiet – women and the poor – he had to employ many techniques to increase participation. 'He listened and prodded them on and used elastic questions, asking back questions origi-nally directed at him. In one review, when a problem arose, the members looked at him for an answer. He laughed, and retorted, amazed at how they expected him, an outsider, to know how best to solve their problem in an area they knew best.' (Mbetu 1997)

The project adopted a 'participatory process approach' which relies on a number of key principles to steer projects, and which even in this planning phase had been applied extensively.

- *Participation* of community members in the decision making and planning as well as the implementation of activities was central to the project. This had been considered necessary for community control of the process. The community was encouraged to make plans, take choices, implement and review them, with as little outside assistance as possible. A truly participative approach works towards the inclusion of all marginalized groups in the community, in particular women and the resource poor, fostering an environ-ment in which they are able to take an active part in decision making.

- *Institution strengthening* is a key contributing factor to participation, as one study of participation in the Chivi project concluded (Farrelly 1995). The project's approach was based on the assumption that, if the work is to be sustainable beyond the period of outside intervention, local institutions must be strengthened to maintain and develop the process after the outsiders have left. The Chivi project has focused on working with and strengthening existing institutions, rather than creating a project structure of its own, in order to further this sustainability. Increasing the confidence and capacity of local institutions is also considered to contribute to their ability to communi-cate effectively their needs and priorities to service providers such as agricul-tural extension services.

  Institution strengthening activities have also formed a significant part of the project's strategy for responding to the needs of the poor and marginalized, especially women.

- *Local skills and knowledge* have been considered the basis for any solutions to community needs. A key feature of the project's approach has been the unlocking of the knowledge and skills already held in the community. For example, at a garden groups workshop, organic pest control measures were shared by older members (see chapter 7). These methods had almost died out in response to the rise of chemical pest control, but are now being explored and evaluated again by community members.

- *Facilitation* ITDG has played the role of facilitator in the process, which has involved facilitating the finding of solutions to prioritized needs from within the community or outside, rather than being the source of those solutions.

This process inevitably takes time as it must follow the community's own pace. It also involves a long period of preparation during which the NGO is learning from community members, rather than transferring skills and knowledge to them.

- *Participatory Technology Development* (PTD) is an approach which could be said to summarize the above points. PTD focuses on increasing technology choice and improving technical capacity. Through strengthening local institutions, building on local knowledge, and facilitating community choice of technical solutions from a range of options, the project has aimed to increase the capacity (technical and otherwise) of community members, both as individuals and in their institutions (the concept of PTD is explored in more detail in chapter 13).

These principles are referred to throughout this book. The process followed can be broken down into a number of key steps, summarized in Figure 5.2. The process of helping the community to list its priorities as described in this chapter are referred to in the fourth and fifth boxes in the figure. The precise timing of each step is outlined in appendix 1: Project chronology.

Time to listen

There is nothing new in the participatory approach, and in fact writers such as Robert Chambers (1983) have been criticizing 'development tourism' for almost 20 years, whereby consultants fly in from abroad hoping to gather information about the project area in as short a time as possible before returning. It is still the case, however, that projects often fail to take time to understand the communities they want to work with.

In southern Africa, and indeed in other parts of Africa, extension work and efforts to develop the rural areas have been continuing for over a century, and yet development has not taken place; indeed these efforts have arguably resulted in greater degrees of dependency than self-reliance. Communities, governments and even CBOs are known to have accepted rather than refused funding for activities that they may not be well equipped to implement, or which are not exactly compatible with their own goals and objectives.

NGOs and CBOs need to budget for more time for the planning and designing phases of their programmes, and ensure that their funding partners understand situations in underdeveloped countries. They should also ensure that project staff are well versed in rural or community development and that they have the skills that will enhance development for the marginalized. This entails internal human resource development seminars and workshops on programme management, communication and networking. In the next chapter the 'Training for Transformation' course is described which was intended to set right some of the top-down methods used by development

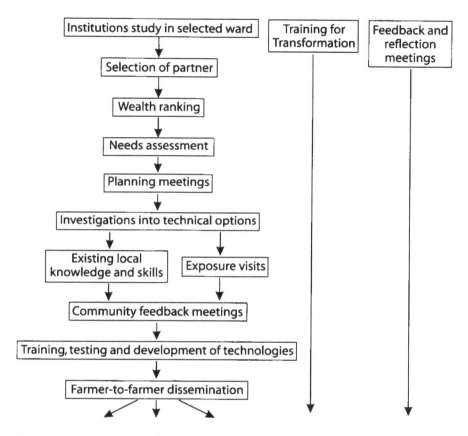

Figure 5.2 Key steps in the participatory process
(Murwira 1996; Watson 1994)

projects in the past by strengthening local institutions and building up the people's analytical and planning skills and empowering them to make decisions about their own lives.

Although the six-month period of investigations and dialogue with the community was crucial for the project in Chivi, it is worth asking whether such a long time is always be necessary in other projects? This depends on the specific situation of different communities, and no general rule can be relied upon. However, when ITDG applied the Chivi process approach to a new project in Nyanga District of Manicaland Province in 1996 some of the components of the approach took less time, partly because of the lessons from Chivi. The project officer commented: 'In Nyanga District we started working together with Agritex from day one. The process is not taking so long, because of the Chivi experience and because we do not have to spend time convincing Agritex.'

6

TRAINING FOR TRANSFORMATION

The need for leadership building

One of the cornerstones of the Chivi Food Security Project was the participation of community members in the 'Training for Transformation' programmes which began in 1992 and continued through to 1995. This training was introduced at the same time as the first exposure visits and technical workshops which are dealt with in the next chapter. Training community members to reflect on their situation and to plan was crucial, since the community in Chivi Ward 21, like the rest of the black population in the country, had been disempowered and deskilled during colonialism. The post-independence era did not bring about much improvement, despite commendable efforts at policy change in favour of the previously disadvantaged black majority. The problem goes far beyond inadequate agricultural policies or poor-quality land, and relates much more to the decline of effective leadership in what had once been self-reliant, inclusive communities.

Changing attitudes towards work parties are a good example of this. Traditionally, any member was able to organize work parties which were attended by nearly every member of the community; more recently, however, the poorest have been excluded from these activities. The work parties brought about the diffusion of ideas, improved production, better harvests and lessened the burden of production processes on individual families. Attendance at the work parties was based on mutual respect and reciprocity. However, colonialism and the introduction of the money market ushered in more commercial behaviour which had a negative impact on social relations. Gone were the times when social relations were based on kinship, mutual respect and age group considerations. Work parties were now being organized on the basis of providing enough food and drink, which meant considerable expenditure on the part of the organizer. Those who could not provide meat and drink, such as the poor in wealth rank 4 (see chapter 4), withdrew their participation from most activities.

In pre-colonial times, the poorest did not go hungry; there were social obligation mechanisms to feed them, mainly through the chiefs' emergency and ritual food supplies, kept back for the needy in difficult times. When there was no emergency, the food supplies would be used in rituals. The effect of colonial rule disempowered traditional leadership, and the chiefs lost their capacity to provide security and food during times of crisis; they also ceased to be custodians of the customs and traditions of the communities, and found themselves unable to provide direction to the community. By 1980 there existed a disempowered, deskilled and resource-poor leadership which could not come to the assistance of its people. Any attempt at introducing a development project which did not address this loss of direction would probably have met with attitudes such as fatalism, a belief in the supernatural, self-blame and a general lack of motivation.

The Training for Transformation Programme

The Chivi Food Security Project aimed to reverse this trend, and 'Training for Transformation' was identified as the most suitable method. Its awareness-raising techniques, based on the psycho-social methods of Paulo Freire, were first developed in Zimbabwe and Kenya by the Catholic Church and were widely used to increase the capacity of groups to identify problems and solutions to them (Wedgwood 1997). The basic training course includes discussions on the meaning of development, principles of transformation, approaches to community development, the psycho-social method and group dynamics. The concepts of self-awareness and self-help are central to this training. The subject matter in the Training for Transformation techniques included:

- What is development and what is the community's vision for development?
- Why are the poor poor? How can they escape from poverty?
- What are the various models for leadership and how can they contribute to development or lack of development?
- How can institutions be built with the potential to transform?
- Group dynamics
- Problem analysis and the identification of solutions
- Communication skills.

The first group of people to be trained at the Silveira Training Centre near Harare in 1992 included the local leadership from the two focus villages in Ward 21, VCWs, VIDCO chairmen, *sabhukus*, farmers' club and garden group leaders. The basic course is in two parts: a two-week session, followed by a break of some months, then a one-week follow-up course. This residential course was found to be ideal for team building, and for challenging the leaders' ideas away from their communities; it was, however, too expensive to send all community members on a residential course. Other community members attended a one-week course held in Chivi itself, which was more accessible, in particular for women and those less able to travel. After the training, one-day summary workshops were held to give feedback to the community.

Most of the community members interviewed said that the training had been extremely useful, particularly in terms of confidence building (see Box 6.1). They were so pleased with the training that they requested that extension workers in the area should undergo the same kind of training. There was also considerable pressure on the Ward 21 councillor to attend training, which he finally did in 1995. The training also provides facilitation skills, which many of those who had been trained later began to use in facilitating community meetings.

Training for Transformation was introduced by the project following the first planning meeting with community leaders and group representatives in Ward 21. However, when the project expanded its activities to a second ward in Chivi, Ward 4, initial awareness-raising meetings based on some of the principles of Training for Transformation were held in each village before the institutions study or the planning meeting took place. This was based on experience in Ward 21, which suggested that the process of initial dialogue with the community would have been greatly facilitated if people had had more opportunity to gain confidence and understand the project process. Following this, the planning meetings progressed more smoothly and there was more shared knowledge about ITDG's aims and approach (Murwira 1996).

Encouraging group representation

Training for Transformation was just the first stage towards strengthening local institutions. Throughout the life of the project, community and group representation have also been encouraged as a method of institution strengthening. During the planning processes which began in 1991, many discussions were held with group and club representatives, and these representatives were selected by their institution and were required to give feedback to the other members on their return from the meetings. This process of selection and feedback, supported by the challenges of Training for Transformation, has aimed to encourage accountability among group members. Groups have been encouraged not to select the office holders to represent them at all meetings, but rather to choose ordinary members, thus increasing the involvement of all members, not just the leaders. For the exposure visits to research stations and other projects (see chapter 7), ITDG requested that no person attend more than one trip, and hence a number of members had the opportunity to see the range of technical options and to report back to the others.

The iterative planning cycle

The project also used an iterative planning cycle (planning–action–review–planning) in order to strengthen the ability of the partner groups, and ultimately the

Box 6.1   Reactions to the Training for Transformation

When you are poor and suffering you forget that you are a person who is worth something. You begin to think of yourself as helpless and you look to other people to come and get you out of the mud. The recurrent periods of drought had killed our spirit. We had also lost the spirit of working together, it was each for himself. The training helped us to look at our problems, our community and ourselves. We studied our groups, their leaders and the problems we had. This training really helped us to understand our problems and to find solutions to the problems ourselves.

Mr S. Masara, farmer, Ward 21

In fact when we can back from the training we found that the Agritex workers in this area needed to go for that training, and we told ITDG to have them trained. Our working relationship improved when the extension workers came back from training.

At first I did not see the need for the training. I had worked in the field for many years – why did I need more training? Yet year after year there was hunger despite all that teaching. Training for Transformation helped me see where I had gone wrong with the farmers. I was not starting from the premise that these were adults who knew the conditions under which they lived. Some of them had been born on the same pieces of land, they knew their fields and understood their lives better than anyone.

Mr N. Mapepa, Agritex officer, Ward 21

Some of us were really changed by what we learned. In the past we saw ourselves as leaders who could not be asked a question. What we said is what we expected to be done. But of course it was just followed badly and people were not happy. We also learnt that we too were experts in our work. This is our village, these are our fields. Yet we were being treated and also acting as if they belonged to someone else. That is why things were not moving. Now in our garden groups and the farmers' groups people are working together in a new way. It is as a result of this training that more people have joined the two groupings as a way of developing themselves.

Mr Madakupfuwa, village development committee member, Ward 21

wider community, to plan, act and review independently. Group representatives plan together, but then feed back to their groups, in order not only to share their decisions, but to have these decisions ratified by the other members.

Groups are encouraged to review their activities, in conjunction with other groups and with community leaders. This process takes several forms. After the introduction of a new technical activity, a mini-review is held, to discuss strengths and weaknesses and make adjustments to plans. A community review is held on an annual basis, at which group representatives contribute their review of the past year and their plans for the future. Several weeks after the community review, village meetings are held to confirm the plans made by group representatives at the review (see chapter 11 for more details of the review process). Small committees and specific planning groups are also formed to organize, plan and review particular events: for example the Seed Fair which takes place annually is organized by a committee of group representatives, which is also responsible for reviewing the event afterwards.

As a result of this emphasis on community meetings and reviews, there have been a large number of meetings in Ward 21 over the life of the project, with high levels of participation, as shown by Table 6.1.

ITDG's aim has been to initiate this iterative planning cycle in the community, and then to withdraw slowly, so that it is increasingly handed over to local institutions (for example, Ward 21 has registered a CBO, see Box 12.1). The aim of enabling community activities that are independent of ITDG has been helped by increased contacts with other institutions.

Increasing links with service providers

Chiefly through the exposure visits, subsequent training and review processes, community groups have increased their knowledge of, and contact with, service providers such as the government agricultural extension service (Agritex) and research departments. This knowledge has in turn led to an increase in confidence on the part of the community in dealing with service providers and a greater capacity to articulate their requirements. The project has aimed to foster direct links between community groups and the service providers and to avoid mediating the relationships in order that this communication can be continued in a sustainable way.

Conclusion

There is no doubt that Training for Transformation has influenced the success of the Chivi Food Security Project. This has been recognized by Agritex and chapter 12 describes how Agritex is in the process of providing training based on Training for Transformation for all its extension workers throughout the country. This is a considerable achievement in itself; however, providing training is only a start. It is easy enough to pay lip-service to the idea behind Training for Transformation, the empowerment of the marginalized farmer. But for this to be taken to heart and to be implemented may require a restructuring of the organization to allow for greater flexibility on the part of agricultural extension workers to respond to the requests of marginal farmers.

Table 6.1 Participants in meetings and training courses, Ward 21

	Total no of People	Meetings/ Reviews	Training for Transform- ation	Visits to projects	Technical workshops	Seed fairs/ Field days	Perma- culture training
1992 male	52	0	38	12	2	0	0
1992 female	110	0	35	17	58	0	0
1993 male	211	99	19	10	83	150	0
1993 female	391	180	40	0	161		10
1994 male	499	471	26	2	0	852	0
1994 female	488	469	0	13	0		0
1995 male	44	9	24	0	11	250	0
1995 female	47	36	0	0	11		0
Total male	806	579	107	24	96	1252	0
Total female	1036	685	75	30	230		10

7

TECHNOLOGY CHOICES

The selection and adoption of various technical options lies at the centre of the project's practical activities. In order to address the problems that were identified and prioritized at the community meetings held in 1991, and which included inadequate water for crops and vegetables, pest attacks and fencing, new technical options must be explored and old techniques revived. This chapter describes the various ways in which the choice of technologies available to the community members was widened: investigating existing knowledge; exposure visits; training; testing of technologies; farmer-to-farmer dissemination; adoption; and farmer participatory research and experimentation.

Understanding existing knowledge

One of the objectives of the Chivi Food Security Project was to build on (and enhance) the traditional knowledge within the communities. In keeping with this objective, and the ethos that people's knowledge must be respected, it was necessary to find out what the community already knew and to try and extend this knowledge base.

Following the needs assessment and prioritization process the ITDG project officer undertook a survey to audit existing knowledge of both water conservation and pest control. In doing this work he used the following methods:

- semi-structured interviews and discussions with individuals who included: traditional leaders like the *sabhukus*, extension workers, older people and other community members
- discussions at group meetings
- observations in both crop fields and vegetable gardens
- feedback and sharing meetings with community leaders.

The survey revealed a number of issues. First, pests were traditionally controlled using natural methods (see Table 7.1 for examples of these). This explains why farmers would say, 'we never used to have pests'. In some instances farmers used and knew local herbs that could be used to control pests, but these practices had died out following the introduction of 'modern' pesticides and discouragement by extension workers (see Box 7.1). Second, traditional methods of soil and water

Box 7.1 Indigenous technical knowledge

I can say that the old [colonial] system destroyed not only our people but our spirit and knowledge. We were behaving as if we never used to think before these people came. This helplessness got worse even after Independence. There are those who are said to know everything and those who do not know anything. We in the villages like Chomuruvati [the 'dark corner'] are seen as knowing nothing because we are poor and we get drought relief. So those who give us things just treat us like that and we let them do it. ITDG reminded us that there were many good things that we knew as Africans and we went back and we realized that we were indeed clever.

Mr Gunge, farmer, Ward 21

It was evident that the farmers knew so much about their own land and how to manage it. It is just that they had not been given the chance to express and share their knowledge not only with one another but with the 'experts'. I remember at one of the discussion meetings when I asked about traditional pesticides. There was one farmer whose tomatoes were never attacked by red spider-mite. But he could not tell everyone what he was using. He was probably afraid of being laughed at or even being accused of not using 'modern methods'.

Gradually as we talked and everybody felt at ease to share what they knew he told us that he used the sap of a local drought-resistant aloe.

Kuda Murwira, ITDG project officer

conservation, such as shifting agriculture and intercropping, were in themselves effective, but these too had been overtaken and discouraged in recent times. Third, some of the new conservation techniques were in fact contributing to the problems, since they were not being adapted and adopted properly, for example, some of the contour ridges running on a gradient were in fact designed to get water out of the fields rather than into the fields, which is not appropriate for such dry areas. Fourth, some of the modern chemicals have a residual effect and can be quite harmful to humans. Some of the farmers were using these indiscriminately. Fifth, farmers were not exposed to alternative soil- and water-conservation techniques and technologies.

The example given in Box 7.1 of a natural pesticide demonstrates that some farmers still used traditional methods of pest control with success, but were reluctant to talk about them for fear of victimization and ridicule. The process of interviews, discussions, observations and feedback encouraged such farmers to share their knowledge with the rest of the community.

Table 7.1   Examples of indigenous pesticides as reported by villagers of Ward 21

Plant (local name)	Application	Control against which pests
Murunjurunju (a climber growing wild in most parts of Chivi District)	Several metres of the plant stem are crushed and mixed with water; this mixture is left overnight, then the solution is applied to leafy vegetables. Care must be taken to avoid contact with skin: users must wear gloves. A period of not less than 7 days should elapse between application date and harvest.	Aphids; bugrada
Machacha/ majacha (a plant growing in heavily manured soils) Leaves used for a relish; fruit used for pesticide	The fruit must be picked when it is yellow and ripe. It is then crushed and mixed with water in a container which should be covered. The mixture is allowed to ferment for two days before use. Sprayed vegetables must not be harvested until 7 days from the day of spraying.	Aphids
Mutsviri/ Mutovhoti/ Muhoza ashes All three trees are commonly found in Chivi District and are normally used for firewood	The ashes of these trees are applied in the planting holes to control cutworms, and on the leaves as dust to control aphids and grasshoppers. In the grain stores, the ashes are mixed with cow dung and smeared on both the floor and walls to control attack by weevils.	Aphids; grasshoppers; weevils; cutworms
Chillies and pepper	Chillies or pepper are planted round the garden to repel pests. Chillies are also crushed and mixed with water; the solution is then sprayed on leafy vegetables. There is no harm in consuming vegetables immediately after spraying.	

Note   These pesticides were reported to ITDG, but their effectiveness and toxicity levels and effects of residues are as yet untested.

Box 7.2 Fencing the vegetable gardens

The need to fence the vegetable gardens was identified in the initial phase of needs assessment and prioritized by the women's garden group representatives. Livestock were destroying the vegetables which were insufficiently protected by brushwood fences. This problem was not addressed, however, in the early exposure visits. After some time, the issue was raised again.

A workshop was held to discuss the issue, attended by two representatives from each of the garden groups in the two focus villages. The workshop debated all the options that the participants knew of, with their advantages and disadvantages: brushwood fences; wire netting; hedges (including leucaena). Finally, a decision was made to follow up wire netting. Four representatives were selected to visit neighbouring towns and price the netting, to feed back to the others and to make suggestions as to a suitable vendor. This took place, and each garden group then calculated how much money they needed to raise to fence their garden, and began collecting subscriptions from their members.

Thus, the problem was identified, various solutions investigated, a particular course of action decided upon and carried out. No financial or other support was given, hence the whole process can be followed again and again for different technical problems, directed and implemented by the groups themselves.

Subsequently, the groups investigated the idea of making the fencing themselves, which led to training in the use of a fence-making machine, which halved the fencing costs (see section in chapter 8).

Similarly intercropping was practised extensively in Chivi in the past but has been discouraged by the extension services in recent years. During the discussions on existing knowledge (supported by information given at one of the training courses described later), farmers regained the confidence in their own skills to reintroduce this technique as a pest-management measure. In this case, increasing technical choice took place within the community, with no need for the introduction of external technologies.

Exposure visits

Often the solutions to priority needs were not found within the community. For example, many of the people in the community had not witnessed such severe droughts in their lives and wanted to explore other communities' ways of coping. When this happened, 'exposure visits' were organized to sources of knowledge outside the community, undertaken by community representatives. This meant that rather than the project officer presenting a single, or even multiple, solutions to the community, its own representatives saw for themselves a range of options, and community members made an informed choice about which to pursue. It

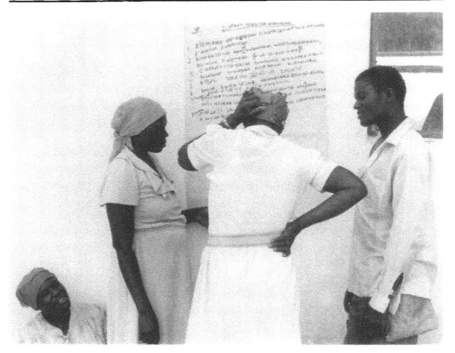

Photo 7.1   Farmers share their knowledge of farming techniques
ITDG/Kuda Murwira

was anticipated that this would give community ownership of the decision – encouraging both motivation to try out a technical option and the freedom to abandon it if it proved unsuccessful.

The project officer identified a number of institutions and locations where possible solutions might be found (including government agricultural research stations, other NGO projects and innovative farmers in other areas). In February 1992, representatives from the two farmers' clubs and two garden groups which had been chosen to pilot project activities were selected by group members to visit the sites. Representatives were chosen carefully to ensure that there was gender equity across the groups; that both leaders and non-leaders of clubs and farming groups were included; that no one individual was allowed to visit twice; that both literate and non-literate people participated; and that the clubs and groups had to reach a consensus on who was to go on a trip. The participation of both men and women allowed the women's needs – in particular with regard to the vegetable gardens – to be taken into account in the choice of options to pursue. For example, the first groups to take part in this process chose tied ridges and sub-surface irrigation: two responses to water conservation appropriate for field cropping and gardens respectively.

Box 7.3 Learning new things through travelling

In Karanga we say *chitsva chiri murutsoka*. This means you see and learn new things through travelling. Some of us had never been outside of Chivi. We were born here and got married here. I went to Chiredzi. My friend I can not even describe to you what I saw. I can show you what I have done in my field. Come let me show you the new things we learnt.

Mrs Wonekayi, farmer, Ward 21

The way in which people were chosen (to go on the exposure visits) was new. We all agreed as clubs and farming groups who should. Even 'frogs' like us went to some of the places. This was good. It is not like in the past where only those who could read and write or those who were master farmers were invited to courses. This made me very happy.

Mr N. Mutonono, farmer, Ward 21

It was interesting to note that when presenting their experiences on the visits, those who had not wanted to go because they were illiterate, were in fact more articulate and made interesting and lively presentations.

Kuda Murwira, ITDG project officer

The exposure visits are also marked the beginning of the direct links which were forged between farmers and the sources of further knowledge, without an intermediary. This, it is anticipated, enables a sustainable relationship to be developed, notwithstanding the need for transport, between farmers and researchers in particular. Increased capacity and confidence in this area, it is hoped, improves farmers' ability to seek solutions from a variety of sources.

Chiredzi Research Station, Masvingo Province

At Chiredzi the farmers saw sub-surface irrigation techniques using clay pipes or pots, tied ridges and live fencing using leucaena.

One woman reported that the clay pipes had made their work so much easier as the technology reduced the number of times they watered their gardens. She said that clay pipes had reduced watering from three to four times to once a week. On top of that they harvested more vegetables after the adoption of the techniques.

Another technique that most of the farmers liked was the tied ridge. Farmers also saw an ox-drawn ridger, but they felt that it would be too heavy for their cattle. Another ridger which was demonstrated had been purchased from a shop and would have been suitable but it was too expensive for them.

The Makoholi Research Station, Masvingo Province

The group that went to Makoholi Research Station saw soil- and water-conservation techniques such as tied ridges and animal traction. They talked to the researchers and learned that it was better to keep cows and not heifers, as cows could reproduce, provide milk and work as much as heifers. They also learned of the importance of feeding their crop residue to cattle.

The farmers learned the benefits of tied ridging, as well as five different techniques of soil and water conservation, some of which proved unsuitable for their conditions and lack of mechanization. Some of the farmers tried out the equipment designed by the Conservation and Tillage (Contill) project based at the research station. These included the disc ridge, ripper tine, tie-maker and some single animal equipment.

Fambidzanai Training Centre

The visit to Fambidzanai Training Centre confirmed the farmers in their use of traditional methods of pest control such as intercropping. The visitors saw natural pest-control techniques, intercropping as a way of controlling pests and also improving yields, tree conservation, beekeeping, textile design and brick-making projects. The farmers were excited but also confused about what they learned; they were glad to see 'traditional' methods of pest control in use and to discover that Fambidzanai actually ran courses to encourage this. They were confused about intercropping, however, as the agricultural extension workers often discouraged them from this practice on the grounds that the beds would look untidy.

The farmers were also very interested in two techniques at Fambidzanai aimed at conserving soil moisture content. The first was the inversion of bottles full of water into the vegetable beds: water is released slowly through a narrow spout over a few days. The second technique was that of spreading plastic sheeting at the base of the bed, then adding up to 60cm of soil. The sheet is 1m wide, and any excess water can run off beyond the edges of the sheet. This helps to retain the water and nutrients within the bed.

Mutoko project

Some parts of Mutoko District are similar in climate to Chivi District. The idea of the visit to Mutoko was to expose farmers to communities that were quite similar to theirs, where farmers had found ways of harnessing the environment to sustain themselves. Mutoko is well known in Zimbabwe for its market gardening activities which supply many parts of the country with vegetables and different kinds of fruit. The farmers were taken to the farm of a certain Mrs Bhaureni who is a successful farmer who has managed to buy a truck, and all kinds of farming implements (see Box 7.5).

Box 7.4 Reactions to the water-conservation techniques

I liked everything that I saw. I felt so much envy and pain, seeing people making so much from their gardens where we were not. I said to myself, surely we would also do it – make money from our gardens. When we came back, I shared with the others what we had seen, and we implemented some of the things such as the *zvikari* (clay pipes for watering our gardens). Before we went to Chiredzi, we could see, but we were partially blind. Now we see more clearly. When we came back home we started making these *zvikari* for our own use. It was easy since most of us knew how to mould clay.

Mrs Anna Gatawa, farmer, Ward 21

These things saved our lives. In the past we only knew contour ridges. We did not even understand why we were digging them. Some people had done them in such a way that they actually drew water away from the field rather than into the field.

Mr Gunge, farmer, Ward 21

Box 7.5 A woman farmer to admire

Most of the people in the visiting group were so shocked. Not only by the amount of productivity they saw but about the fact that here was a successful woman doing it all by herself. She is also the district chairperson for the Zimbabwe Farmers' Union in Mutoko. This was a new experience for most of the visitors as this is quite rare in Chivi.

Kuda Murwira, ITDG project officer

Box 7.6 'How come we never heard of such a thing before?'

When I visited Zvishavane I was so happy with what I saw. This man [Mr Phiri], showed us these pits that he had dug to 'hold the water' in his fields. It looked so simple and I said, how come we never heard of such a simple thing before? That is why in Karanga it is said *kusaziva hufa* (lacking knowledge is as good as being dead).

Mr Madekupfuwa, village leader, Ward 21

Photo 7.2 Tied ridges was one of the most popular techniques for conserving water

ITDG

At Mutoko the visitors were exposed to soil- and water-conservation techniques such as tied ridging. The farmers were also able to compare the quality of crops that were planted on tied ridges and those that were not. They saw that the ridges made a great difference.

The farmers from Chivi also met members of a savings club who saved to raise money for buying farming inputs. This 'co-operation' impressed most of them, as this was an issue that was still problematic in Chivi. The visitors were able to hear how farmers pooled the little resources they had and in this way were able to raise a lot more money than they would have done as individuals.

Zvishavane water projects

The visit to Zvishavane water projects exposed the farmers to more water-conservation techniques. Like Mutoko, Zvishavane is a very dry area. Farmers working with the Zvishavane water projects have learned methods of water conservation, water harvesting and moisture retention. This is what the visitors from Chivi witnessed. All those who went were particularly impressed by what they saw being practised by other peasant farmers.

After all the visits to the research stations, the community in villages C and E met to get feedback from those who had gone. The report back sessions were very lively. Some of the farmers drew on the ground what they had seen and how it worked. Those who had had the opportunity to see comparative situations of before and after the technologies were adopted were able to share with the others how significant was the difference the technologies made. Each technique was described and weighed in great detail; how easily adaptable it was, how much it cost, how heavy, how much input was needed materially or financially.

This process was facilitated by the community members who had gone on the visits and not by the ITDG officers. This enabled those who had gone to express what they had seen in their own way and analyse it themselves. In this way it was not the ITDG staff who were describing each technique and trying to persuade the community to take it up, but the community members discussing with one another.

After a few of the technologies seen had been described and discussed, the community reached a consensus on which ones were best suited to their environment or could be adapted, and were affordable. The community then decided that the on-site trials of the technologies that had been chosen should begin in two group gardens and two farmers' clubs. In addition, some individual farmers were anxious to put into practice what they had seen straight away.

Training

When the technical options had been selected by the community meetings, training was arranged for the groups and clubs who were to pilot the process, facilitated by ITDG. In most cases, the training took place in the villages, led in the initial stages by outsiders from the relevant research station or project (later training was led by community members – see chapter 12). In one instance, that of clay pipe-making for sub-surface irrigation in the vegetable gardens, training for the women's garden groups was carried out by women from Chiredzi town, who had been working with the local research station staff on clay pipes and had been producing the pipes themselves for sale in the town. The technologies included were: sub-surface irrigation; pest management; water harvesting; crop diversification; and shallow well improvement.

Testing, adaptation and analysis of technologies

Following training, in 1992 and 1993 farmers started experimenting with the new options. The testing and adaptation of technologies by farmers themselves aimed to increase their technical capacity and strengthen their ability and confidence to experiment and innovate. Some carried out their own comparative trials, planting, for example, on top of the tied ridges in half the field, and on the side of the ridge in the other half (see later section on farmer experimentation). Various modifications were made by individual farmers on their own initiative. Mini-reviews were held during and at the end of the season to facilitate the sharing of the results of this stage, at which farmers shared their experiences, innovations and analysis with each other.

Farmer-to-farmer dissemination and the seed fairs

Dissemination of appropriate solutions within the project has focused on the process of 'farmer-to-farmer exchange'. This is based on the principle that farmers, both men and women, are best placed to share with each other the results of new ideas and innovations, as they usually share the same values and understanding and have trust in each other's analysis. This process of sharing information takes place irrespective of outside intervention, and in the case of Chivi is evident in the spread of techniques from villages C and E to other villages in the ward via friends and families (see Box 7.7). However, the project has also attempted to foster dissemination further through various activities. The methods used in both the exposure visits and the training, whereby group representatives were responsible for sharing information about new options with fellow members, aimed to encourage farmer-to-farmer exchange. Non-participants in the testing of new technologies were invited to the mini-reviews, to enable them to be exposed directly to the new ideas.

The community has also adopted the agricultural extension service's concept of field days and competitions and modified it into a community-led process, whereby officials are invited by the community to act as judges and given the criteria for judging. The field days enable farmers from the surrounding area to

---

Box 7.7   The adoption of technologies

Ever since we started using these new skills we had learned we have not stopped. We use *zvikari*, or the bottles, we plant many varieties of vegetables and we mix them in the beds, and our families now have enough vegetables to eat and we even sell some. We do not 'break our necks' fetching lots of water for watering anymore. We water once a week and that is enough.

Mrs Chiza, garden group member, Ward 21

We know that *Njere moto unogokwa* [knowledge is like a fire of which you can ask for a live log from someone]. So we are going to copy this knowledge from our relatives right here in this ward.

Mr Dzviti of village D, who learned the new techniques by observing what was happening in neighbouring villages

Photo 7.3   Farmers displaying their crops at a seed fair
ITDG

see for themselves the new ideas, and to talk to the farmers who are implementing them. Similarly, the annual seed fair, at which farmers compete, individually and as clubs, for awards for the greatest number of varieties grown and for the best produce, facilitates the sharing of information and ideas.

The idea of the seed fairs arose out of the problem of suitable dryland crop varieties for the climate in Chivi. The objective of the fair was to revive local seed varieties, share information on these and acknowledge that these crops do thrive in Chivi. At the seed fair, which is like a trade fair, farmers display local seeds, for example, mhunga, millet, rapoko and so on. The farmer with the best-looking varieties is awarded a prize. These fairs have been held since 1993 and each year get bigger and better; they are planned and organized by the community members, who set the judging criteria themselves and also contribute the prize money. This has boosted the farmers' confidence and increased the sense of ownership of their own activities. In the past, shows and field days were judged according to criteria set by extension workers or outsiders, which were not always the same as the farmers'.

The results of the seed fairs is now apparent. In most of the fields in Ward 21 farmers now grow small grains and a wide variety of drought-resistant crops which are doing very well. Seed fairs are now an annual event and are held at ward level. Again this has helped the farmers to learn from one another and also enhanced co-operation among the community members.

Adoption

After testing and dissemination, the new techniques and technologies were adopted by farmers and gardeners. However, this is not a linear process, and adaptation and experimentation did not therefore cease at this point. An iterative process of trial and error, sharing of ideas and modifications, continues as a part of the food production cycle within the community.

Throughout this technology development phase, the agricultural extension service, Agritex, was closely involved, joining in the training sessions, participating in the mini-reviews and learning not only about the technical results but also about the process of participatory technology development and the capacity of the farmers to manage that process themselves.

Farmer participation in research and experimentation

A number of technologies, new to the Chivi villagers but known elsewhere, were adopted when the villagers returned from their exposure visits. These, and the indigenous technologies rediscovered through workshops, are described and their uptake documented in the next chapter. As well as the dissemination of existing techniques, however, the project also aimed to influence the development of new technologies and genetic resources.

Historically, agricultural research and extension policy in Zimbabwe has not encouraged farmer experimentation. Research has generally been based around on-station trials, and the results passed to the extension services for dissemination to farmers, usually in the form of blanket recommendations with an insistence on rigid adherence to the message. As a result, adoption rates have been low (Hagmann et al. 1995). Internationally, there is now increasing recognition that this model of extension is irrelevant, if not contrary, to the needs of poor farmers (Scoones and Thompson 1994).

Part of the aim of the project has been to make and develop links between farmers and researchers, in order both to share the results of research more directly with farmers and not just with the extension services, and also to influence the research agenda to make it more responsive to the needs of farmers. Research, if it is to be effective, needs to be informed by farmers' priorities and in particular their criteria for success:

> In this [transfer of technology] mode, priorities are determined by scientists .... In the new, complementary ['farmer first'] mode, this process is stood on its head. Instead of starting with the knowledge, problems, analysis and priorities of scientists, it starts with the knowledge, problems, analysis and priorities of farmers and farm families.
>
> (Chambers et al. 1989)

For example, in addition to the factors of yield and drought tolerance which are usually the primary criteria for research, farmers often place an importance on reliability, taste, storage value, ease of processing and pest resistance. These priorities reflect their multiple role as producers and processors of food, as cooks and consumers, in contrast with traditional researcher's single perspective.

Through the exposure visits and subsequent training sessions, research station staff have been able to make direct contact with some of the farmers in Chivi, and a relationship is developing which has the potential to continue beyond the life of the project. ITDG has been collaborating with the GTZ-funded Contill project based at Makoholi Research Station to make the case to Agritex and the Department of Research and Specialist Services for a more participatory approach to research and experimentation. ITDG is also recommending that Agritex organize district level fora for extension and research staff to meet with farmers to discuss the research agenda and share findings, rather than the current provincial level meetings at which farmers are not represented. The role of the extension worker is also affected by these proposals: they can play a key part in encouraging experimentation and facilitating the spread of information from farmer to farmer.

In addition to the testing and adaptation of technologies and techniques described earlier in this chapter, other types of trials are taking place in Chivi as a result of the encouragement of farmer experimentation and research. First, the research station staff are taking some of their trials to farm sites in the district and working with the farmers. The area committee of farmers' clubs in Ward 21 facilitated a meeting at which two farmers from each village were selected to carry out the trials. This has been an interesting process, as initially the researchers dictated to the farmers exactly how the trials should be conducted, with the result that the farmers' non-trial crops performed much better than the trial plots (i.e. without the restrictions and rigid guidelines laid down by the researchers). The farmers then persuaded the researchers to allow them to conduct the trials in their own way, using their own knowledge of when to plant and so on, with far more successful results. The trials included different crop varieties (maize, sorghum and cotton) and techniques (e.g. moisture conservation and fertility management). Post-season workshops enable a review of the results and the management of the trials (Watson 1993b; ITDG 1993c, 1994c).

The second type of trial taking place in Chivi is initiated by the farmers themselves. For example, in response to farmers' requests for information on suitable dryland crop varieties, the project officer obtained two millet varieties from Matabeleland. The farmers' clubs' area committee organized the distribution of the seed to farmers to try out in their fields. Farmers have also been testing cassava and castor beans (ITDG 1993d, 1993f, 1995c). ITDG has facilitated and supported this process, the results of which have been as much in the increased confidence in farmers to conduct their own experiments, as in the outcome of the trials themselves.

8

TECHNICAL OUTCOMES

The exposure to new technologies, re-emergence of traditional techniques and encouragement of farmer-participatory research described in the last chapter resulted, as one observer put it, in an 'explosion of experimentation by local farmers, turning fields into plots for testing new ideas for soil and water conservation and for planting techniques' (Mulvany et al. 1995). This section summarizes the main technical activities undertaken in Ward 21 as a result of the processes described above. The activities are presented here according to the needs to which they were responding. The initial planning workshop identified the following priority needs:

Field crops

- water for crop production
- animal draught power and equipment
- suitable dryland crop varieties.

Vegetable gardens

- water for vegetable production
- fencing for protection from livestock
- pest management.

Both sectors

- combating increasing landlessness
- community co-operation (ITDG 1993d).

Details of the impact of these activities are presented in chapter 9, although a brief indication of the uptake is given here.

Water for crop production

As water conservation was identified as the priority need for both vegetable and crop production in the initial stages of the project, there has been the greatest

Table 8.1   Water conservation techniques for field crops

Technology	Source	Uptake 1991–7 (approx. no. of households)[1]
Tied ridges (see Figure 8.1)	Chiredzi and Makoholi Research Stations; Harare Institute of Agricultural Engineering at Mutoko	450
Infiltration pits (see Figure 8.2)	Farmer innovation (Mr Phiri from Zvishavane)	850
Rock catchment	Farmer innovation (Mr Phiri from Zvishavane)	14
Modified contour ridges	Farmer designed	40
Mulching and ripping[2]	Makoholi Research Station (Contill Project)	5
Fanya-juu terraces (see Figure 8.3)	Contill Project	4
Winter ploughing	Revived traditional practice	800
Intercropping	Revived traditional practice	450
Termite soil as fertilizer and moisture retainer	Revived traditional practice	800

1. Total number of households in the project area approximately 1300.
2. Technique involving 'ripping' into stubble with an animal-drawn ripper and planting in the riplines.

variety of activities in this area, as Table 8.1 shows. The technologies and techniques selected by the farmers' clubs for testing, in this and indeed the other technical areas, focused on the whole on low-input, low-investment activities, which the farmers deemed most appropriate to their circumstances. They contrast in many cases with the high investment (in time and money) and high-risk recommendations advocated by agricultural extension staff.

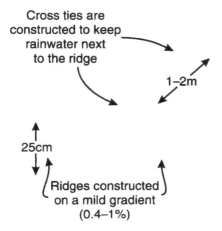

Figure 8.1 Tied ridges retain water; planting is then carried out into a fully moist ridge

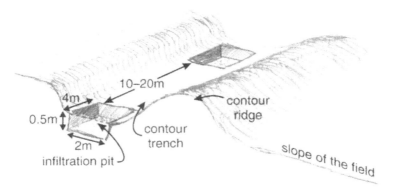

Figure 8.2 Infiltration pits are dug in the channel next to contour ridges to trap run-off

Animal draught power and equipment

Trials have been carried out with tillage implements designed by the Contill project based at Makoholi Research Station, such as the disc ridger, ripper tine, tie-maker, and some single animal equipment. However, the main constraint in animal draught lies in the availability of livestock for ploughing and land preparation, in particular since the 1990–91 drought, in which livestock losses were heavy. Activities in this area have therefore centred around increased co-operation within the community in sharing animals during the ploughing

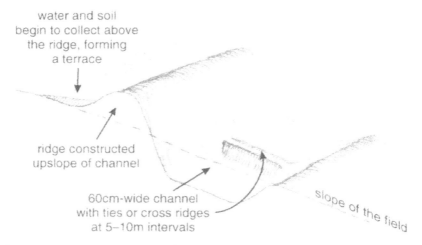

water and soil
begin to collect above
the ridge, forming
a terrace

ridge constructed
upslope of channel

60cm-wide channel
with ties or cross ridges
at 5–10m intervals

slope of the field

Figure 8.3 *Fanya-juu* (Swahili for 'throw up-slope')

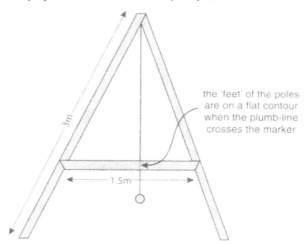

the 'feet' of the poles
are on a flat contour
when the plumb-line
crosses the marker

3m

1.5m

Figure 8.4 The 'A' frame, constructed from wooden poles, string and a stone, made marking the contour much easier

season; working together in club groups on each others' land to prepare fields by hand where necessary; and investigating the possibility of accessing government loans for restocking with the help of the ZFU.

Suitable dryland crop varieties

The exploration of suitable dryland crop varieties has focused on three activities. The first is the creation of an annual seed fair, at which farmers display, as individuals and as clubs, the range of field crop varieties they have grown. Prizes are awarded for the greatest diversity between species, as well as for the best quality produce. This is a key opportunity for farmers in the community and from the surrounding area to identify local sources of seed and exchange information on their performance. In the vegetable gardens, demonstrations and field day competitions have performed a similar role for vegetable crops.

Second, there have been trials of a number of different crops and varieties carried out by farmers in Chivi District (see chapter 8). Up to 140 farmers were involved in the cassava trials, while 100 tested the castor beans.

Third, 30 farmers (one representative from each of the farmers' clubs in Ward 21) have been taking part in a seed-bulking programme initiated by GTZ–SADC. The farmers are given one sorghum and one millet variety, plus training in seed selection and crop breeding. The aim of the programme is to strengthen local seed production and develop local capacity in this area.

Water conservation for garden vegetables

Table 8.2   Water conservation techniques for garden vegetables

Technology	Source	Uptake 1991–5 (approx. no. of households)*
Inverted bottles (see Figure 8.5)	Fambidzanai organic training centre	300
Mulching	Fambidzanai organic training centre	800
Clay pipes for sub-surface irrigation (see Figure 8.6)	Chiredzi Research Station	450 women
Infiltration pits	Farmer innovation (Mr Phiri from Zvishavane)	7 gardens
Gully reclamation	Farmer innovation (Mr Phiri from Zvishavane)	2 gardens + 4 villages
Plastic sheets buried in vegetable bed	Fambidzanai organic training centre	250
Shallow well improvement	Zvishavane Water Projects	8 gardens

*   Total number of households in project area approximately 1300.

Figure 8.5 Upturned bottles direct water straight to the root zone, reducing evapotranspiration. The moist soil acts as a 'plug', reducing the rate at which water drains from the bottle. This technique is suitable for fruit trees and leafy vegetables with a row spacing of at least 30cm

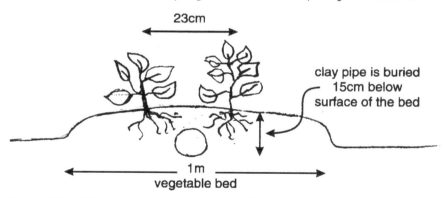

Figure 8.6 Sub-surface irrigation using locally made clay pipes

Fencing the gardens

Protecting the vegetable crop from roaming livestock during the dry season (when forage is scarce) is a major problem for gardeners. The existing fences of brushwood are often breached, and require heavy maintenance; they also carry a high environmental cost.

The first part of the process of tackling this problem has been described above (see Box 7.2). Following the choice of wire netting, two men were selected to go to

Photo 8.1   Fencing to keep out animals was seen as a priority in the garden
groups and is now made locally

ITDG

Silveira House Training Centre in Harare to learn how to make both the wire
netting and the machine which constructs it. On their return they brought the
two machines with them and manufactured fencing for the garden groups, who
contributed the plain wire and some labour costs (ITDG 1993d). This fencing
cost half the price of ready-made fencing; the fence makers have also trained
others in their village in the operation of the machine. By mid–1993, five out of
the 40 groups in the ward had completed fencing their gardens (ITDG 1995b).

A second response to this problem has been the development of live fencing
using sisal, a traditional fencing material. This has been taken up by the garden
groups who have less money available to buy the wire fencing; by mid–1993,
eight groups had taken up this idea (ibid.).

Photo 8.2   Women using a 'form' to make clay pipes

ITDG/ Kuda Murwira

Pest management

Garden group members in Ward 21 have been involved in four activities to improve pest management. The first is the use of local plant materials as pesticides, based on traditional local knowledge supplemented by training given by the Fambidzanai Organic Training Centre in Harare. The training has helped to revive local knowledge and has encouraged the validation and sharing of existing skills in the community, particularly by the older members. By mid–1993, more than ten of the 40 garden groups in Ward 21 were using traditional and biological techniques for pest management (ITDG 1994b). Investigations are under way to find an institution that can test the toxicity of indigenous plant pesticides, to increase knowledge of and control over this resource.

Similarly, the technique of intercropping, which was previously practised but abandoned in the face of disapproval from the extension services, has been revived as a pest control measure in the gardens as well as the fields. The training courses given by Fambidzanai staff identified a third technique: the use of repellent plants in pest control. This has subsequently been introduced.

Finally, gardeners identified the need for training in the correct use of chemical pesticides. Those who could afford it were at times using chemical pesticides,

> **Box 8.1** A new problem – marketing
>
> Garden produce is in abundance and we have problems disposing it. We have to dry a lot of the vegetables that we grow. Because of transport problems, we wait for customers to call on us and also risk losing them. We sell on credit but some people never pay their accounts. Cotton sales are better than garden sales. One major problem is that we do not have a place of our own where we can sell our produce freely. Customers are dictating prices because we have not matured enough along these lines. Wherever we go to customers to sell our produce, customers peg prices for us.
>
> *Farmer, Ward 21*
>
> Once we have a local market, we intend to invite buyers so they can buy directly. We have already set up committees who are going to spearhead the marketing of produce. These committees will start marketing as soon as we start ploughing so that by the time we harvest, we should have already secured markets.
>
> *Farmer, Ward 21*
>
> People wanting to sell [vegetables] have had to travel as far as Ngundu and Chivi service centre and transport to those areas is expensive. We would like the Minister of Lands and Agriculture to look into the issue.
>
> *Farmer, Ward 21*
>
> *Source:* interviews with one of the authors

but were aware that they did not know the correct dosage or application methods and were concerned to eliminate possible danger and increase the efficiency of their use. A training course was subsequently run by staff from Chiredzi Research Station.

Vegetable marketing

The need for improvements in the marketing of vegetables did not arise until the second and third year of the project, as garden production improved in response to the influx of new ideas and enthusiasm. One of the responses to this issue was to examine ways of diversifying production. As a result, demonstrations were held in the gardens in Ward 21 on sewing a variety of new crops: carrots, beetroot, spinach, eggplant, green pepper, chillies, lettuce and potatoes. Group representatives bought seed for trials of these new crops, which were then tested in most of the gardens in the ward.

Photo 8.3   Clay pipes for sub-surface irrigation reduced the need for watering
by about half

ITDG/Kuda Murwira

Representatives from the garden groups also visited Masvingo town, to
explore marketing options. At supermarkets, hotels and markets they investigated which products were available, price levels, and prospective demand. They
fed back to a workshop attended by gardeners, following which each garden
group made plans for which vegetables to grow in the coming season (Watson
1993a).

Increased co-operation and increasing landlessness

In addition to the priority needs listed above, both garden groups and farmers'
clubs identified two further concerns: the need for increased co-operation, and
increasing landlessness. The former is being addressed by activities such as the
Training for Transformation courses and the participatory planning and implementation process which increases the capacity and co-operation of local institutions (see chapter 5). There are no activities directly addressing the problem of
increasing landlessness, although efforts to improve food production per unit
area aim to alleviate land pressure, as does intercropping through more efficient
use of available land. The Training for Transformation courses also attempt to
contribute to the democratization of community decision making, which has the
potential to encourage more equitable allocation of available land, although there
is insufficient data at present to evaluate this.

9
IMPACT

Introduction

The Chivi project faced a particular challenge in assessing impact within the participatory process, without counteracting its aims of fostering only activities initiated and led by the community. In response to this challenge, the project has begun a new participatory monitoring system, which is described in chapter 11. This chapter focuses on the evidence available at present on the impact of the project's activities, placing particular importance on the community's own perceptions of benefits and disadvantages. Impact is measured against the three main objectives defined by the community below.

During a workshop in November 1995, at which a new monitoring system was finalized (see chapter 11), a list of objectives for the project was drawn up, in priority order, and indicators identified for each objective. The three main objectives, which were considered of equal importance, were as follows:

- to strengthen co-operation
- to share skills and knowledge
- to strengthen household food security (ITDG 1996; the full list of objectives and indicators is presented in Table 11.1).

Objective 1: Strengthening co-operation

Growth in membership of institutions

The impact of project activities on the farmers' clubs and garden groups in Ward 21 has been dramatic. The number of farmers' clubs has doubled, from 17 to 33, as have the number of garden groups, from 19 to 41. Overall membership has risen from 324 to 989 for the farmers' clubs; and from 635 to 1200 for the garden groups (the total ward population was approximately 184 households in 1990). This increase in membership is not solely the result of more clubs and groups forming. Membership within individual institutions has increased: for example, Nyanyira garden group increased from 18 to 23 members, and Bati garden group from 22 to 27 members between 1992 and 1993, and at the same time, two farmers'

Box 9.1 Working together is important

After independence we had a slogan which went *pamberi nekubatana* [forward with unity]. We used it very often in this ward. But it did not mean that we were united. We said it but did not do it. Each person just did their own thing at their home. The greatest thing that ITDG has taught us is co-operation.

Mrs Gwatiziva, farmer, Ward 21

When ITDG came to us, we were not united. Those without anything could not work with those with cattle and ploughs. Master farmers had their own groups which those of us without a donkey could not join. A few women had gardens which the rest did not want to join. The only other place where we used to meet was at ZANU (PF) meetings. But we could not really say we were co-operating. So when some of us went to Mutoko and told us about the savings clubs, those who went to Chiredzi told us about the garden groups and how they were working, we realized we were oppressing ourselves.

The work we do in this area is very hard, particularly if you have no draught power, if you do not have a plough, and other things. So we realized that working together was important.

Mr Masara, farmer, Ward 21

clubs increased their membership to include all the residents of the *sabhukus* village. At a farmers' club workshop in 1993, only one of the 23 clubs represented had not experienced an increase in membership. Growth in club and group membership is assumed to be a positive indicator of a perceived benefit by those joining or remaining in the groups, and hence an indicator of the positive effect of the project. However, the project may also have benefited in part from a revival in these institutions which was taking place just as the project began.

As a result of the increased involvement in these local institutions, the level of participation in the project's activities is very high. The study of participation in July 1995 estimated that about 88 per cent of households in the ward were participating in the project – around 1150 households, out of an estimated 1300 (all membership figures in this section are from Farrelly 1995).

As well as reaching many (breadth of outreach), the farmers' groups and garden groups are benefiting the poor (depth of outreach, see Box 9.2). In Ward 21 in 1997, 80 per cent of households falling into wealth rank 4 were reported to have been assisted with carrying out at least one of the following operations by their fellow club members: ploughing, threshing millets, digging and spreading anthill soil and digging infiltration pits. During these operations the less fortunate were given access to draught power owned by their club members as well as manpower they would have found very difficult to obtain otherwise.

Accountable leadership

The accountability of group leaders to their membership is a topic discussed as part of the Training for Transformation course, as participants are encouraged to assess both their own leadership qualities and those of other leaders. ITDG has encouraged this process, in particular focusing on greater involvement of the whole community, including the more marginalized, in group activities and decision making. As a result of these activities, there is evidence of changes in the leadership and operation of the farmers' clubs and garden groups. Meetings are being held more often and more regularly. More elections are taking place (historically, post-holders had failed to call regular elections according to the institutions' constitutions and had remained in post for many years).

In a survey of 20 groups (both farmers' clubs and garden groups) whose leaders had been exposed to Training for Transformation, one group claimed that their leadership was excellent, 13 said theirs was good, five said satisfactory and one said unsatisfactory (Hagmann, Chuma and Murwira 1995). At a leaders' workshop in Ward 21 in late 1994, community leaders summarized the impact of the Training for Transformation courses thus:

- empowerment of women in decision making
- improved co-operation in communal activities and at household level
- understanding that people have equal responsibility irrespective of sex
- empowerment of people to do things they thought impossible
- sharing information (ITDG 1994e).

Increased ability for self-reflection

During a Training for Transformation course in May 1995 for traditional leaders, attended by 24 of the 33 *sabhukus* in Ward 21, the participants identified problems in their own approach. These included poor responsibility for common property resources; a failure to motivate people for self-reliance; failing to call regular meetings; and dictatorial tendencies (Farrelly 1995). This increased ability for self-assessment is considered an important requirement by the project if local institutions are to become more capable of responding to the community's needs.

Organizational ability

There is evidence that the organizational ability of the groups in Ward 21 has also increased. At the beginning of the project, staff took responsibility for arranging workshops, meetings and project visits, albeit with as much consultation as possible. For some time now, however, this role has been taken on by local institutions (ITDG 1995b). For example, as mentioned earlier, the annual seed fair is organized by a small committee chosen for that purpose, made up of a representative from each village. They decide on the date and venue, organize collections for the prizes, invite local officials to be the judges, and review the event afterwards.

In 1992, the community hosted a visit by 150 members of Oxfam-supported projects from around the province, a visit they co-ordinated themselves, as they have succeeding visits. Also in 1992, women from the two focus villages made their own independent arrangements for training in shallow-well improvement by staff from Zvishavane water projects, with no assistance from the project. In more recent times, the community has planned further exposure visits, to neighbouring areas Mwenezi and Chisumbanje, and contacted the government extension service to request the use of the provincial bus. At a workshop organized in October 1995 by the ward councillor for Ward 21, participants (including area committee members, representatives from farmers' clubs and garden groups, *sabhukus*, VIDCO chairs and other local leaders) summarized some of the achievements of the project as follows:

- We are now able to organize our own meetings like shows, seed fairs and field days without assistance from outside;
- We are able to arrange our own elections for area committees for both farmers' clubs and garden groups;
- We are able to encourage participation of community members to join farmers' clubs and garden groups;
- We are able to host visitors from different parts of the world and share with them our experiences;
- We have linked with different institutions to enhance development in the area, e.g. Agritex and the Research Department;
- We have followed up promises made by outsiders to ensure they are fulfilled, e.g. ZFU, the District Administrator and Agritex (ITDG 1995d).

Increase in confidence

Closely linked to increased organizational ability is an increase in confidence, a vital factor in institution building, which has been noted among the groups in Ward 21 (ITDG 1995b; Wedgwood 1996a). This enables groups to enter into dialogue with service providers, not only at local level (for example, the ward-based agricultural extension worker) but also at district and provincial level, through participation at meetings, discussions during officials' visits to the ward, and so on.

The benefits of effective, representative institutions have been recognized by the community: in November 1992, following a workshop run by Zvishavane water projects on institutional development, the groups involved in the project decided to form an overall committee at village level, to enable them to co-ordinate their activities better between the various farmers' clubs and garden groups. Consequently, two committees were formed, one in each of the focus villages, composed of the chairs of all the garden groups and farmers' clubs in that village. ITDG had no involvement in this process at all.

Elections

Having organized a representative, co-ordinating body at village level, community members then began to be increasingly discontent with the farmers' clubs' area committee (the ward-level institution which represented them in the ZFU hierarchy), on the basis that it was unrepresentative and ineffective. The ZFU constitution required this committee to have seven members, all of whom should be affiliated to the ZFU. The existing committee was not drawn from all the villages in the ward and had not held elections for many years. Eventually, it was agreed within the ward to hold fresh elections, and agreement was sought and gained from the ZFU district chair that the committee could consist instead of 12 members, two to be elected from each village, and that affiliation to ZFU was no longer a prerequisite, but could take place afterwards. The election was held in June 1994 and since then the committee has been active in helping to co-ordinate activities in the ward.

At the same time, the garden groups began to express concern that there was nobody representing their views at ward level. Following discussions, it was decided to elect a similar area committee for the garden groups. ZFU agreed that this body could be considered as a 'commodity group' (a specialist committee) and as such have official ZFU recognition and be able to send representatives to district level ZFU fora. Accordingly, representatives of all the garden groups in the ward met to elect a committee made up of two representatives from each village. The two area committees meet on a regular basis, both separately and together, to discuss and plan activities in the ward.

Box 9.3   Community fundraising

In 1996 project workers helped Ward 21 Area Committees to write a funding proposal which was submitted to HIVOS, a development agency from the Netherlands. HIVOS agreed to initially fund the training component of the proposal, at a cost of Z$300 000. A committee of area chairpersons (of farmers' clubs and garden groups) reporting directly to HIVOS, now manage the project finances, prioritize training needs, identify and commission trainers and evaluate the outcomes. So far courses have been held in vegetable processing and preservation, budgeting, business management, marketing, blacksmithing, tree planting, alternative water pumping and leadership.

Participation by women in local institutions

There has been a noticeable increase in the involvement of women in local institutions. They have always been prominent within the garden groups, but are under-represented in leadership roles in farmers' clubs (where membership is approximately 60 per cent men, 40 per cent women). Because of the parallel involvement of women's garden groups alongside farmers' clubs in the project processes, women have equal representation in bodies such as the village overall committees described above; indeed the chair of one of these committees, and the vice-chair of the other, are women. There are also a considerable number of women who have undergone Training for Transformation, some of whom have become key facilitators at workshops and meetings such as the annual community review.

Project staff have noted an increased confidence in women to speak out in meetings, challenging others' (including men's) point of view, and speaking after their male counterparts (which goes against tradition) (Murwira 1996). This is mirrored by a shift in intra-household relationships described in chapter 10.

Finally, there has been a discernible impact on the relationships between the various institutions in Ward 21 and the surrounding area. In particular, the conflict between the *sabhukus* and VIDCOs has lessened, due in part to the discussion about their roles which the Training for Transformation course initiated with their leaders. The *sabhukus* are regaining their position in the management of natural resources (ITDG 1995b). The ward councillor, who resisted Training for Transformation for several years despite pressure from his constituents, has attended and, according to community members, now has better relationships within the ward as a result.

Box 9.4   Revolving loan funds in Ward 4

Funds have been established by members of a garden group in Ward 4 to help them save money for purchasing assets they would otherwise have found diffi-cult, if not impossible, to acquire on their own. Each month the members contribute a small amount of money into the fund; the total amount is then given to members in rotation, to buy large items. Four members of Hamamaoko group have received Z$400 each. One bought a sewing machine, one acquired an ox-cart, and the other two bought six goats each.

Two members of Budiriro used the funds to buy cement and roofing sheets for their homes. One bought a wheelbarrow and a door, another bought chicks for rearing. The activities in these two groups is representative of revolving loan funds in 19 of the 25 garden groups in Ward 4.

Impact on inter-household relations

Farmers are providing more support to each other not only through communal labour, but also through lending of assets, for example, cattle for ploughing, donkey carts, tools and other equipment. In addition, the community's ability to identify and prioritize needs has increased, both as a group and as individuals. Facilitation skills have been gained by many community members, and this has enabled those who were not necessarily leaders before, and in particular women, to play a leadership role in facilitating meetings and workshops.

The 1996 evaluation report concluded that participation has developed insti-tutional capacity among farmer groups in Ward 21 in the following ways:

- An increase in confidence and the ability to make demands on local service providers, notably the Agritex extension worker;
- An increased network of contacts facilitated by the project, which farmers are now making use of independently of ITDG;
- Group cohesion that encourages joint work and problem solving;
- Increased capacities of groups to plan, to seek solutions to problems and to manage change.

Objective 2: Sharing skills and knowledge

According to the participants at the monitoring workshop, the activities of the project have increased levels of knowledge among farmers and gardeners, and perhaps more importantly have increased their confidence in their existing knowledge and skills, and their ability to innovate for themselves (ITDG 1996). Evidence of the increase in and sharing of skills and knowledge can be seen in the

Box 9.5 A model of good practice

Mr Haruzivi, a farmer in Ward 21, won the 1997 Good Farming Competition. He shared the secrets of his success at a field day, held at his homestead. An elaborate system of soil- and water-conservation techniques allowed him to intensify production on his field, by planting a month earlier, and growing two maize crops in one season. Because of the residual soil moisture he managed to increase his sorghum and maize plant population by 12 per cent, and also grow vegetables such as tomatoes, kale and cabbage. He also planted rice in contour ridge channels, producing 180kg from two strips, 2m wide and 100m long.

Photo 9.1 Mr Haruzivi shares his water-conservation techniques

ITDG

> **Box 9.6** A definition of food security
>
> If we see children playing outside very happily, mother and father sitting together and not quarrelling, and the dogs eating some left-overs, we will know ah, this home has enough food.
>
> *Suggested indicator for food security at a community meeting*

uptake of technical activities. As a result largely of the participatory processes described in chapter 7, the technologies introduced have spread from the pilot groups throughout Ward 21 and beyond. For example, 50 farmers tested tied ridges and infiltration pits in the first year; in the second year over 200 were implementing them. Similarly, 62 women began using clay pipes, now over 400 do so (Mulvany et al. 1995).

There are also numerous examples of farmers from outside the ward, and in some cases beyond the district boundary, coming to visit and learn from those in Ward 21, and taking home the new techniques and technologies to try out (Murwira 1996).

As well as the informal sharing of ideas, as the Chivi project began to expand from Ward 21 to Ward 4 in 1994, and then to other wards in Chivi and Mwenezi District in 1998, experienced farmers began to act as resource persons who could facilitate training sessions. In 1998, ten farmers from Ward 21 organized a training for 30 representatives from ten farmers' clubs in Ward 22, and received Z$16 000 for their efforts. The 30 representatives were then expected to report back to their farmers' clubs and pass on their training.

Objective 3: Strengthening household food security

At the monitoring workshop, the first two objectives, strengthening co-operation and sharing skills and knowledge, were considered prerequisites for the third, strengthening household food security. This latter has also been the overall aim of the project since its inception. The workshop participants identified indicators for the objective of food security as follows:

- soil- and water-conservation activities
- the availability of suitable crop varieties; and
- the presence (and preservation) of crop surplus.

These indicators are used here to consider the impact of the project on household food security. In addition, the indicators identified for the objective of generating income (that is, improved vegetable production and the production of cash crops) are also considered.

Photo 9.2   Sub-surface irrigation with clay pots helps women get better yields
ITDG

Soil- and water-conservation activities

The uptake of technologies in this area has been widespread (see Tables 8.1 and 8.2). Water conservation in both fields and gardens was identified as the top priority need in the original discussions, and soil- and water-conservation techniques have, from the beginning, excited the most interest and enthusiasm (see Box 9.5, which describes one farmer's production of two crops in one season as a result of soil and water conservation). In 1995, it was estimated that approximately 80 per cent of households in Ward 21 were practising soil- and water-conservation techniques in one form or another (Hagmann et al. 1995).

Tied ridges and infiltration pits continue to be taken up by farmers, encouraged by the positive results of their neighbours. At a visit by Chiredzi Research Station and Agritex staff to the project in late 1992, the farmers explained that

Photo 9.3   Infiltration pits were widely adopted, in spite of the initial labour
required

ITDG

they found tied ridges hard work but effective in harvesting water and conserving soil, and that crops planted on the ridges grow faster than those in plots without ridges (ITDG 1995a). The end of season review in 1993 concluded that tied ridges and infiltration pits are efficient methods of conserving soil and moisture, and that tied ridges help to minimize the problems of draught power and labour shortage in land preparation. In addition, there is also a positive effect on soil erosion in the area (ibid. 1992e).

Disadvantages of these measures include the heavy labour input – in particular tied ridges and infiltration pits – (ibid. 1992e); and the danger of water- logging in heavy rains, although the ridges are designed so that the ties can easily be broken to allow excess water to flow away (ibid. 1992e, 1993e). The work required to construct the tied ridges and infiltration pits is considered, on the whole, as capital investment, as the levels of maintenance needed are considerably lower than the initial construction work (ibid. 1995a).

According to the women gardeners, the clay pipes laid in the vegetable beds reduce watering by around 50 per cent on average, varying with soil type and weather conditions (Mulvany *et al.* 1995). The main constraints lie in the manufacture of further pipes: on seeing the positive results in the pilot gardens, other garden groups wanted to acquire pipes, either through purchase or manufacture by themselves. However, there have been difficulties in obtaining the correct design of form, which has led to poor quality pipes being produced. As a result of this, and the labour input required to manufacture the pipes, interest has moved from the clay pipes to plastic sheets and inverted bottles, although those who have pipes continue to use them.

Availability of suitable crop varieties

The number of varieties and species of field crops in production in Ward 21 has increased every year, based on the samples on display at the annual seed fair (ITDG 1994b). Farmers are trying out new varieties, both improved and traditional, and sharing their knowledge with each other. This increased access to and knowledge of new varieties is seen in the plans made at the community reviews: for example at the 1994 community review, farmers groups made specific plans to grow an increasing number of crops (ibid. 1994a).

The range of crops produced in the vegetable gardens is also increasing, partly encouraged by problems of marketing the increased surplus production. When the project first began, the common crops grown were cabbage, rape, onions and tomatoes. Now new crops such as carrots, beetroot, spinach, eggplant, green pepper, chillies, lettuce and potatoes are being tried out (ibid. 1993b).

Presence of crop surplus

There is evidence of both increased yield and improved reliability of cropping in the project area. Although precise yield data is difficult to collect, as it is considered private information, one farmer reported that he planted sorghum in three roughly equal-sized plots, one with no ridges and two with tied ridges, planting one side of the ridge in one plot, and on both sides in the other. The plot with no ridges yielded 30kg; that with ridges planted on one side 80kg; and that with ridges planted on both sides also 80kg (ibid). There are further indicators of improved yield and crop reliability in the reported increasing presence of buyers' trucks, which come into the area to buy the harvest of ground nuts, sunflower, and so on (Murwira 1996).

The disadvantages of increased crop production are felt largely by the women, who are responsible, in the main, for crop processing. Particularly at times of abundant harvest, the women are often constrained from beginning the necessary preparations for vegetable production in their garden plots.

Box 9.7 Marketing a crop surplus

Ward 21 area committee representatives met with agricultural produce buyers, who agreed to set up buying points within the ward. In 1997 they bought 30 tonnes of unshelled groundnuts worth more than Z$78 000. Agritex records for 1997 show that more than 80 per cent of the smallholder farmers in the two wards managed to sell some crops. The closeness of the produce market to the smallholders allowed even those with very small quantities (one bucketful or more) to carry their produce to the selling points on their heads and dispose of them at a fair price. Women generated substantial cash income from the surplus groundnuts. Mr Gatahwa of Ward 4 said he was speaking on behalf of many men in his community when he said that next year he is going to depart from tradition and grow his own crop of groundnuts, a crop normally reserved for women.

Photo 9.4 Cultivating maize on the tied ridges

ITDG

Photo 9.5   Different local seed varieties attract a lot of attention

ITDG/Margaret Waller

In the vegetable gardens, the emergence of the problem of marketing is an indicator of increased production, in terms of both the number of people producing vegetables and increased yields (ITDG 1995a). Most households are now producing enough for their own consumption, so the surplus must be sold elsewhere. Women are sending their husbands to neighbouring areas on their bicycles (which women do not ride) to sell their produce – a new phenomenon (ibid. 1994c).

Improved vegetable production

As mentioned above, there have been increased sales of surplus vegetables. This has resulted in greater availability of cash, evidenced by the ability of most garden groups to collect sufficient funds to buy fencing; and by the establishment of a rotating fund in one of the gardens (each member contributes a sum monthly and the total is given to members in turn). There are instances of women from neighbouring wards coming to rent one or two beds in the Ward 21 gardens, in order to gain from the knowledge and activities taking place there.

Production of cash crops

The absence of cash-crop production was one of the criteria for the selection of Ward 21 at the beginning of the project, as cash crops usually require some cash investment (in seed, inputs, etc.). However, cotton production has increased in the ward since 1990 (Murwira 1996), although it is not yet clear which house-holds, from which wealth ranks, are involved in this activity.

The fertility and biological pest control measures being practised in Ward 21 contribute to production at lower cost than purchased inputs of fertilizer or pesticides, thus reducing margins, and are also more easily available.

10
GENDER ISSUES

The gender context in Chivi District

Women make up about 70 per cent of Zimbabwe's rural population (Murwira 1994). Female-headed households are common, some of whom are widows and some are women whose husbands are working in the urban centres. In a sample studied in Ward 21 of Chivi, 16 per cent of all households were female-headed (Farrelly 1995). They are generally the poorest households, in particular the widows, and are least likely to participate in institutions such as the farmers' clubs or garden groups.

Membership of the garden groups in Chivi is about 90 per cent women, whereas men form around 60 per cent of the farmers' club members. As discussed earlier, leadership of the garden groups is nearly all by women, but they are under-represented in the committees of the farmers' clubs. Other local institutions in Ward 21 are similarly male-dominated: for example, the WADCO, and to a lesser extent the VIDCOs, are made up largely of men; *sabhukus* are always men; and all 29 wards in Chivi District are represented by male councillors.

Women are generally responsible for vegetable production in Chivi, while men take most of the decisions relating to field crops, although women are actually involved in almost all of the crop production processes. Particular tasks such as weeding and winnowing are primarily performed by women and post-harvest processing is usually carried out exclusively by women. Men's role in the vegetable gardens is confined on the whole to site clearance, fencing and some digging.

Traditionally, women had responsibility for particular field crops such as groundnuts. However, this role is beginning to change, as male unemployment encourages men to consider groundnuts as an alternative source of income. This reflects the difference in priorities according to gender: men's primary concern is generally cash crops, while women's is subsistence production. Turning to ownership of assets, larger assets such as ploughs, tools and equipment, are owned by men; whereas women own small household items, pots and pans and so on. Income from crop sales and livestock production is controlled in general by men, while women have access to income from surplus vegetables and chicken sales, in addition to any income from selling crafts (Lewis 1996).

Changes in gender relations

Women's control over natural resources and assets has not always been as limited as it is in present day Zimbabwe. In the pre-colonial era, there was a clear division of labour, as there is today, but the roles of men and women were more complementary, with men and women each owning land and cattle.

In Shona and Ndebele custom, women obtained land-use rights through their membership in particular patrilineages. Male lineage heads obtained land from chiefs and headmen and then allocated this land within their subsistence units. Women were allocated land-use rights in their capacities as wives and daughters in patrilineages. Married women were allocated land for use by their husbands. There was also a family field to which the husband, wife and children contributed labour. Produce from the family field was used for entertaining visitors, to pay tribute, or for consumption when the woman's food from her portion ran out. Daughters obtained land from their fathers and they grew crops for food which could be exchanged for other property in readiness for marriage. Divorced daughters could also look after themselves by working this land assigned to them by their fathers (in other words, through the agnatic patrilineages).

As opportunities for wage labour appeared in the colonial era, men often had to stay away from home, leaving women to provide all the labour on the farms. Married women were increasingly becoming *de facto* managers of households and making decisions over production, especially where the husbands worked in distant towns without good communication with the rural areas. At the same time women's control over the assets of farming – land and livestock – was curtailed by the introduction of legislation. Women lost their rights to land through the 1951 NHLA which recognized men as the registered landholders. Only registered landholders were allocated grazing land, which meant that women could not have their cattle registered in their own names. In cases of litigation or destocking to satisfy agronomic criteria, the livestock of the women were likely to be assumed to be owned by the men in whose names they were registered (Gaidzanwa 1988). To that extent, the NLHA contributed to the low status of women and present-day problems relating to women's entitlement to land and property ownership.

After colonization, competition between men and women for land set in. Land for grazing was allocated to married men, and women were entitled to land only if their husbands had deserted them or lived outside the country. Young adolescent women, who would previously have been assigned land by their fathers, were increasingly left out in situations of land scarcity.

Relations within the household were affected by the appearance of money as a means of exchange and since then the tendency has been for men to carry out activities of greater monetary value. The conflict surrounding the division of labour has largely been created by wage labour. When activities that were traditionally done by women became more profitable, men tended to grab the opportunity and the activity either became a job or developed into a small-scale industry. Women used to do the cooking, but when catering began to provide paid labour, men took over. Activities that have evolved like this include cooking, grinding, and horticulture. Most grinding mills in Chivi are owned by men. Women traditionally grew vegetables and had developed preservation and seed-selection techniques, but when vegetables became cash crops and the technology improved, the activity became reserved for men.

As men took control of most of the new income-generating activities – whether labouring for a wage or growing cash crops – women's remaining work in subsistence agriculture was considered to be of less importance, and their position in the home less powerful. At the same time, men have also had to experience redundancies from paid employment together with the uncertainties and the loss of a role that this brings.

Men's attitudes towards the empowerment of women are determined by their view towards power relations within the household. In pre-colonial times, men's roles within the household sanctioned their power over women in the public domain. In the post-colonial period, when men are often away from home, and women manage the farms, any ideas of empowering women that come from government or development agencies may be viewed as a threat by men. They cannot see how the changes will affect their family lives, and they fear that women's roles as providers of physical and emotional support will be eroded.

The project's gender strategy

The project's key principle of participation has led it to focus on raising awareness within partner institutions about the importance and benefits of full community participation, including all marginalized groups. Gender is a key area of inequity among these groups.

Training for Transformation has been a major vehicle for awareness raising on this subject, supplemented by discussions and facilitation by project staff, which have continued throughout the life of the project. The fundamental principle behind it is that change will be more sustainable if driven by the community members themselves (Murwira 1996). If the project were to focus on widows, for example, it is possible that this would actually decrease community co-operation and responsibility.

Box 10.1   Introducing a gender strategy without confrontation

So if we had gone there and said that one of our objectives was gender, nobody would have listened to us. Sensitivity to custom led us to employ participatory rural appraisal techniques to achieve equal opportunities for men and women. The tools of analysis facilitated changes in gender roles. For example, the garden groups were mainly for women, but men began to participate in garden projects. Women started to accept leadership roles, they also began to be vocal at meetings.

Kuda Murwira, ITDG project officer

This definite but subtle strategy is complemented by other activities which positively encourage the participation of women in the project, notably through the garden groups. As outlined in chapter 4, the garden groups were deliberately selected as a partner institution in order to facilitate the involvement of women. As a result of this, equal numbers of men and women were usually included in the selection of representatives, for instance for exposure visits, and for participation at reviews, as there are roughly equal numbers of garden groups and farmers' clubs involved.

In addition to the above, there have been a number of specific gender-orientated activities in the project. In early 1995, some representatives from Chivi attended a 'women in development' training course at Silveira House. In February 1996, following questions raised at one of the project reviews, at which community representatives were present, a gender workshop was held in Ward 21 at the request of the community, to discuss gender issues. A gender study was also carried out in late 1994, which revealed some useful baseline data, but did not provide in-depth analysis of the gender issues in the project (ITDG 1994f). Consequently, a second study was planned for late 1996, to include a measure of gender training and awareness raising for project staff and community members. This aspect of the study was intended to clarify the topic of gender in the project, to attempt to avoid the issue becoming simply a response to outside pressure with little comprehension of the reasons behind it, or commitment to it.

Impact of the gender strategy

The women involved in the project are generating more income for themselves than before, largely through increased vegetable production. They are now in a position where their husbands ask them for money, rather than vice versa (see Box 10.2).

Box 10.2 Women's increased income and confidence: some comments

Now I don't need to ask my husband for money.

Men working in towns knew their wives would come from the village for money. Now they wait for their wives to bring vegetables.

Before, the husband would tie a string around the money in his pocket so that the wife would not hear that he had money. Now we share our money.

Before the project, it was not easy to nominate another man's wife or someone else's husband to a leadership position in case questions were asked as to how you knew that person's competencies and skills. Also those in leadership positions are now aware that they cannot stay there for ever, that we can vote them back in or out.

Box 10.3 Women's savings clubs

In Zimbabwe, women have had a long history of raising modest amounts of money in a savings club that can be used by the club members whenever emergencies arise, or whenever economic opportunities emerge. This was partly a result of the fact that under colonial rule women were considered minors, which meant that any legal arrangements were done on their behalf by their fathers, husbands, brothers or sons. Property was not registered in their names but in the names of their menfolk. To that extent, the question of access to bank loans did not even arise, and women have had to be resourceful in order to get hold of urgently needed cash from time to time.

These savings clubs are known by a number of different names, and the way in which money is raised varies. The savings club fund is not kept in a bank, usually because of the long distances to the nearest bank, the inconvenience of accessing the fund or the illiteracy of the club members. Each time the fund is collected and allocated to one of the members, it is used almost immediately to buy much-needed food supplies, household equipment, small livestock such as goats and sheep or even luxuries like bicycles or wardrobes. If a crisis arises at some other time, a woman could immediately sell her goats or chickens to gain access to cash. Most of these fundraising activities are informal and are based on mutual respect.

In Chivi some of the women have started a loan fund for members. The non-interest loan fund enabled individual members to meet their own specific needs. Most of the women have bought ploughs, goats, donkeys and even cattle. Some informal groups agreed on what household items to buy for each member, such as pots, plates and cutlery. They continued buying until each member had received them before deciding on the next need. It is possible that the idea of the loan fund arose out of the visit to Mutoko where the farmers from Chivi were introduced to members of a savings club.

Photo 10.1 Mrs Wonekayi receives a prize for her vegetable garden from the ward councillor

ITDG/Margaret Waller

Photo 10.2 The project manager joins a women's dancing group at the seed fair

ITDG/Margaret Waller

Some of the positive impacts of the project's gender strategy and activities are summarized below:

- The women in the garden groups have increased income at their disposal, which is changing the balance of power within the household, as the quotations in Box 10.2 show. At the same time, they are increasing in confidence in debates and discussions for community decision making; are taking facilitatory roles in community meetings; are attending training courses, some away from home; and are taking up more leadership positions – for example, the chair of one overall committee and the vice-chair of the other are both women (Lewis 1996; see also ITDG 1994e).
- In the farmers' clubs, women are taking a more active role, as members and in decision making, but they continue to be absent from the key positions, such as chair, at present.
- The creation of the area committee for garden groups at ward level (see chapter 9) is a significant step in gaining status for the key women's activity of vegetable production. Through this committee, women can influence the farmers' clubs area committee. However, the latter committee only has one woman member at present, and so in itself represents largely the male farmers in the clubs.
- At the household level the vegetable gardens are also attracting more notice, in part because they are making a more significant contribution to household income, particularly in drought years when the field crops fail. This has had a positive impact on the status of the work, both within the community and among agricultural extension agents.

Box 10.4 Men growing women's traditional crops

In 1997 many women in Ward 21 were able to take their surpluses to marketing points set up in the ward, and they earned significant cash income from the sale of groundnuts. Their success has interested a number of men in growing groundnuts, although this has traditionally been considered a woman's crop.

The idea of men growing groundnuts for sale may be seen as an indicator of the new respect accorded to women's economic activities. However, since men tend to own more land than women it may be that cultivating groundnuts will be effectively taken over by men, and that women's success in this farming operation will soon be forgotten. History is full of examples of the marginalization of women once an activity is commercialized. When activities that were traditionally done by women became more profitable, men tended to turn the activity into a job, or a small business. Grinding food crops used to be done by women at home, but now most grinding mills in Chivi are owned by men.

Box 10.5 Strengths and weaknesses of the gender strategy

Strengths

There has been a positive impact, as described above, on women's access to the means of production, on their involvement in community decision making, and in the distribution of household income.

By supporting an activity in which the women were already involved as part of their current role (that is, the garden groups), these benefits have not added enormously to women's labour burden. There are, however, repercussions for women's labour when the field crop harvest is large and gardening activities are delayed because of longer crop processing periods.

Again, building on current gender roles in food production has meant that existing knowledge and skills have been developed in both women and men, and reduced the potential differential between men and women on the introduction of completely 'new' technologies and processes.

The process of debate and dialogue around gender issues allows the subject to be discussed with all members of the community, and contributes towards a sustainable change, rather than potentially alienating sections of the community by an explicit focus.

Weaknesses

There are some disadvantages under which women labour, which the project has not yet tackled directly. For example, women are less mobile, because of domestic commitments and cultural constraints, and are therefore not as free as men to attend meetings and training courses, etc. away from home. Some of the meeting schedules are inconvenient to women's timetables.

Women are still under-represented in both membership and leadership of farmers' clubs, although they continue to be directly involved in field crop production. This is a difficult area where more work is needed.

The farmers' club issue represents a wider concern as to whether women's 'strategic' needs are being met or avoided through the project.

Project staff have suggested that a gender study should have been carried out at the beginning of the project, to provide useful baseline information, which could have formed the basis for discussion in the community.

The disaggregation of monitoring data by gender has only taken place since 1993, hence historic analysis of project activities by gender is difficult.

11

UNDERSTANDING THE PROCESS

The case for re-orienting monitoring and evaluation

Monitoring and evaluation are usually carried out by government departments, implementation agents and NGOs for their own purposes as well as for satisfying the requirements of funding agencies. Evaluation reports are meant to inform the NGOs or agencies and are not designed with the project participants in mind. The process of gathering data therefore tends to be extractive: those who collect the data, and those from whom it is extracted, may not understand the reasons behind the system, or the use to which the information is put. Collation and analysis often takes place away from the project site, and the conclusions may not necessarily be fed back to the project participants.

NGOs' and funding partners' requirements for monitoring and evaluation are understandable: there is a need to keep track of expenditure as well as the progress being made. However, for ITDG as a support and facilitating agency to impose the collection of data for external use on the Chivi project would clearly undermine the participatory process. For sustainability and growth, it was crucial that the community be involved in evaluation and that they be empowered to carry out evaluations which enable them to make relevant and appropriate decisions about their activities.

In designing a monitoring and evaluation system for the Chivi project, the challenge lay in finding a means whereby the necessary information could be collected and made accessible to various stakeholders without undermining the participatory process. The role of the community in controlling the collection of this information was vital, as access to knowledge was the key feature of the project's approach. The original monitoring system met with varying success, and so in early 1996 it was replaced by a more participatory system.

This chapter describes how the original monitoring system worked, and how the new monitoring system built upon it.

The original monitoring system

Once the main objectives of the project had been agreed upon, the monitoring system was designed to measure them based on three key elements: a monitoring file, mini-reviews, and annual reviews. The monitoring file was simply sub-divided into the three main project objectives, with a list of indicators for each objective (drawn up by project staff) at the front of each section. Project staff made regular notes of activities and observations under the relevant heading in the file. This information was summarized on a yearly basis under three topics: exposure to a particular activity, uptake of that activity, and impact.

Mini-reviews were held as a regular part of the iterative process adopted by the project. Several months after a new technology had been introduced (or an existing technique shared more widely), a mini-review workshop was held to enable those who were testing the ideas to discuss them with others. These workshops facilitated farmer-to-farmer sharing of information and encouraged further adoption and adaptation of new techniques and ideas. The mini-review workshops were tools for self-evaluation by the community in a particular technical area.

Community reviews were held annually to discuss the objectives and aims of the project, a review of the previous year's activities, and plans for the following year. These plans were taken back to village meetings for sharing and confirma-tion. Participants included representatives from all the farmers' clubs and garden groups in the ward, and local traditional and formal leadership representatives. Increasingly as the project progressed, the community reviews were facilitated by community members themselves, using the skills they had learned from the Training for Transformation courses.

ITDG also held an annual project review as part of its internal planning, monitoring and evaluation system. This usually took place after the community review, in order that it could be based on the findings of that meeting, but also took into account ITDG staff's role in the project and the wider considerations of replication. Community representatives and other external stakeholders also attended this review.

The weaknesses of the original monitoring system originated mostly with the monitoring file. Although indicators, once developed, were shared with the community and discussed at the community review and other meetings, they were not owned or prioritized by them. Farrelly's study of participation highlighted the differences between community-generated indicators and the project staff's view. In contrast to the original 28 indicators – an unmanageable number – selected by the project team, a group of farmers and gardeners came up with three: the existence of good vegetables and gardens; being taught the importance of working together without excluding anyone; and the absence of anthills: these are used as a soil conditioner on the fields, but require a high level of co-operation as hard work is involved (Farrelly 1995). The information summarized from the monitoring file was used largely by ITDG and other stakeholders, rather than by the community. There was an emphasis in the data collected, on quantitative rather than qualitative information, and a focus on activities carried out, rather than impact. Finally, there was no disaggregation of data by gender or wealth rank. The community review, although a successful mechanism for reviewing and planning activities in a way which is increasingly controlled and organized by community representatives, has similarly lacked an evaluative emphasis.

The new monitoring and evaluation system

In late 1995 ITDG began a discussion within the community about monitoring. The first step was a discussion with the community leaders at ward level to decide on how the process would be managed. The facilitators from ITDG pointed out the need for the community to set indicators and ways of monitoring their own progress, so that they would be able to identify obstacles and set targets for themselves. At this meeting it was agreed that one farmers' club and one garden group from each of the six villages in the ward would be selected (excluding the original groups which piloted project activities) to discuss monitoring with facilitators from ITDG. At each meeting the groups started by re-stating their objectives as a group and then deciding what would be suitable indicators for each objective.

The results from these discussions were synthesised by ITDG staff and presented to a larger community leadership meeting. After this the participants were split into groups to discuss the objectives that had been identified by the 12 groups and to rank them in order of importance. A list of objectives for the project was drawn up, in priority order, and indicators identified for each objective. The objectives and the indicators are given in Table 11.1.

There was considerable debate at the workshop on the priority order for the first three objectives, with unanimity on the overriding importance of all three. The discussion centred around whether co-operation or skills and knowledge came first; the final decision shows the consensus that good co-operation is necessary for the acquisition of skills and knowledge, and that both of these are requirements for strengthening food security.

The participants at the workshop (representatives from the farmers' clubs, garden groups and community leaders within Ward 21) concluded that most of the objectives were in the process of being fulfilled, with good progress on many. The overall conclusion was that the participants had attained food security to a large extent. Community members defined food security as 'beating hunger' (Scoones and Hakutangwi 1996b).

The Chivi project forms part of ITDG's International Food Production Programme, and as such is also part of the POEMS monitoring system (see Box 11.1). At this meeting a plan was drawn up for a participatory monitoring system in which each farmers' club or garden group secretary would keep a record of activities and impact, based on the agreed indicators, and submit it on a monthly basis to the area committee. The information, it was anticipated, would cover membership, problems and solutions, innovations and visitors. This written system proved

Table 11.1   Farmer-designed monitoring and evaluation

Identified objective	Evaluation indicators
Strengthening co-operation (*kubatana*)	Formation of new groups
	Helping one another with draught power
	Men and women helping one another (without suspicion from the other's partner)
	Being able to solve disputes if they arise
	Organizing shows and fairs
	Sharing information
Sharing knowledge and skills	Sharing information on farming methods, e.g. moisture-retention techniques
	Using appropriate drought-resistant crop varieties
	Going to other areas to observe and learn
	Organizing meetings to share this information
Finding ways of ending hunger	Adopting new water- and soil-conservation techniques such as infiltration pits and tied ridges
	Planting seed and crop varieties that are suitable for the area
	Practising winter ploughing and mulching
	Using natural fertilizers, like anthill soils
Soil and water conservation	Use of ridges
	Winter ploughing and gully reclamation
Enhancing self-reliance	Farming well in both fields and gardens
	Using one's hands, for example, to make crafts for sale
Earning income	Growing crops like castor beans, cotton and sorghum for sale
	Using one's skills to make crafts for sale
Draught power	Growing saleable crops in order to buy and sell cattle and donkeys

Box 11.1   POEMS

The ParticipatOry Evaluation-oriented Monitoring System (POEMS) used by ITDG proposes that monitoring impact primarily means monitoring change, and does not rely on pre-determined indicators. It acknowledges that change is largely unpredictable, but can be grouped in domains. Project workers selected the following domains:

- changes in people's technology choice
- changes in people's livelihoods
- changes in other organizations or institutions as a result of the influence of the project
- other significant changes which happen in the lives of the people in the project which may be in domains completely outside the scope of the project.

These changes are documented by project staff monthly and shared with other stakeholders.

ineffective; many secretaries were semi-literate, and documentation was poor.

The community opted instead to use verbal briefings, whereby representatives share significant changes taking place within their own clubs and groups at ward level meetings held every three months.

The strength of this new system is that it facilitates the community's own review of its activities (more frequently and in a more evaluative way than the annual community review used to), thereby enabling them to control the process and strengthening their own institutional base. At the same time, project staff are continuing to record information which will provide useful baseline data for external purposes, without intruding on or disempowering community members. It is interesting to note the convergence of approaches by the POEMS initiative and the community's system on monitoring change across a broad spectrum of the project's impact, rather than the earlier more mechanistic gathering of information on activity levels.

Although an external evaluation was originally scheduled for mid–1993, this was postponed as it was felt by ITDG that it would interfere with and contradict the process being adopted by the project at that time, and that the existing monitoring system should be sufficient to provide the necessary data. However, the study of participation in the project carried out by a post-graduate student in mid–1995 yielded useful information on involvement in project activities by wealth rank and gender (Farrelly 1995). The results of the study were shared and confirmed by community representatives at feedback meetings.

An external evaluation took place in mid–1996, focusing in particular on the impact of the project, in order to provide useful data for donors and other stake-holders and in particular to make the case for the participative approach. The methodology was participative and project staff endeavoured to prepare community members for the evaluation, making clear its purpose, and ensuring that findings were fed back to community meetings before finalization of the results. The evaluation was carried out by two external consultants, one of whom is a member of the government agricultural extension service (see Scoones and Hakutangwi 1996b).

In addition to the external evaluation, a further study has been undertaken to document impressions of the project from the community's point of view. This was carried out in February 1996 by a communications consultant, who inter-viewed community members about their views on the benefits and lessons learned from the project which they wish to share with others (Win 1996). The consultant also identified people and topics for a planned video, which again enabled the community members to tell the story of the project from their own perspective.

12

BEYOND CHIVI

Chapter 3 described the process by which first a province, then a district, followed by a ward, and finally two villages were selected for the start of the Chivi project. This chapter describes how the project began to have an effect beyond the first two villages, first upon neighbouring villages in the ward, then upon new wards, 4 and 22, and upon an area within a new province of Zimbabwe, Nyanga in Manicaland. Not only was the example of the farmers of Ward 21 influencing an ever wider geographical area, like a ripple in a pond, but the participatory approach to local food security is being copied and institutionalized in Agritex to make the extension department more responsive to the needs of poor farmers. Influencing the government's agricultural extension services to adopt a participative approach to extension was one of the project's original aims (see chapter 2). The Chivi project has also had an impact on other departments within ITDG Zimbabwe, on other NGOs working in Zimbabwe, and on organizations in Southern Africa and further afield.

This chapter considers first the process of replication and information sharing in general, then looks more closely at the strategies followed to influence extension policy in particular.

Within Chivi District

The original plans of the project involved operational work in two wards in Chivi District, to be followed by a second project in another district of Masvingo Province. This strategy, it was anticipated, would produce sufficient evidence to make the case for the participatory approach to food security and agricultural development in Zimbabwe's communal areas.

When work began in the first ward, 21, in Chivi District, the two poorest villages were selected for initial activities. Consequently the needs assessment, wealth ranking, planning meetings, exposure visits and initial training all focused on these two 'focus' villages in particular, with the pilot groups (two farmers' clubs and two garden groups) selected to test the new ideas. Following the initial testing phase, information was shared between these pilot groups and other farmers and gardeners through informal channels, supplemented by 'farmer-to-farmer' training workshops and mini-reviews facilitated by ITDG (1995a).

Box 12.1 Ward 21 registers a community-based organization

By 1998 Ward 21's committee of Area Chairpersons (of farmers' clubs and garden groups) had obtained funding from HIVOS (see Box 9.3), and over 100 people had received training in skills such as poultry-keeping, welding, tie-dyeing and blacksmithing. Trainees have now started community-based businesses in carpentry and poultry-keeping.

In addition to managing this grant, the area committees also received fees for providing trainers in agricultural methods. Farmers from Ward 22 were interested in receiving training in some of the techniques they had observed in Ward 21, and so 10 of the most competent farmers from Ward 21 were selected to teach 30 representatives of ten farmers' clubs in Ward 22. These 30 trainees would then provide training to their clubs, which each comprised about 30 men and women. The joint training demonstrated to the farmers of Ward 21, and to other farmers in the vicinity, the value of the skills they had acquired over the years: not only did other farmers want to copy the techniques, but another institution was willing to pay for training. The trainers from Ward 21 received Z$8000, and Z$8000 went to the area committees.

In order to manage these funds responsibly, Ward 21's area committees are in the process of registering their own CBO, with a constitution, a Board of Trustees comprised of prominent people from the ward and from Chivi District, and a bank account. The funds are at present being used to start up the income-generating activities being initiated by community members. It is particularly important that as ITDG's project staff direct their attention to new wards and regions, Ward 21 builds its capacity to raise and manage funds efficiently, so that they can make the most of their skills as resource persons, and also sustainably control the use of funds raised.

About 18 months later, in mid–1993, a planning meeting was held in the remaining villages in Ward 21, to begin a similar process there. By this time, the communities in those villages were already aware of activities in the focus villages and were keen to initiate a similar process themselves. Similar steps were followed, but instead of exposure visits to external institutions, farmers from the focus villages shared their experiences and provided training where possible (although some external visits were also made). Some farmers from the focus villages also facilitated some of the planning meetings and workshops, having by this stage attended Training for Transformation courses. The agricultural extension officer for Ward 21 was more involved in this spread of the project to the remaining villages in the ward.

As well as this form of 'horizontal' replication within Ward 21, further 'vertical' developments continued within the initial focus villages, in the form of new

Photo 12.1   Planning together is a crucial part of the agricultural calendar
ITDG/Farai Samhungu

technical options to address the prioritized needs, the identification of further needs, and the adaptation of new ideas.

In 1994, work began in the second ward which had been selected by Chivi District authorities, Ward 4. The process was similar to that used in Ward 21, with a few notable differences (Murwira 1996; Vela 1996; Watson 1994): first, village awareness-raising meetings were held in each of the villages in the ward, facilitated by Silveira House training centre staff. Based on the experience of Ward 21, it was decided that such meetings could play an important role in introducing the community to the process and clarifying expectations at an early stage. This meant that the subsequent planning workshop could be much more focused, since it built on a common understanding of participatory development. Second, the agricultural extension worker was closely involved from the very beginning, taking part in the initial studies and dialogue and helping to facilitate explanatory meetings, and the agricultural extension officer and supervisor for the area were also involved. Third, work commenced with all six villages in the ward at once, rather than just two. Each village selected two institutions (a farmers' club and a garden group) to pilot the activities. Finally, rather than using exposure visits to sites outside the district, the selection of technical options in Ward 4 has relied greatly on the farmers of Ward 21, with visits in both directions, and local training courses.

It has always been ITDG's intention to facilitate the process in such a way as to become redundant, its role taken over by a combination of the community and the extension service, Agritex. The starting point for the work in Ward 4 therefore, was to increase Agritex involvement from the very beginning. This was facilitated by the process already taking place in Ward 21, which could be used as an example, and by the fact that there was time before work began in Ward 4 for the extension workers from that and surrounding wards to go on the Training for Transformation course to prepare them for their facilitative role.

In the period 1998–2001 the work in Chivi District is being extended with even greater participation by Agritex extension staff. Five wards have been chosen – Wards 1, 10, 23 and 28 of Chivi District, and Ward 2 of Mwenezi District – with the intention of carrying out projects based on the Chivi experience in each ward. Agricultural extension officers and community members from these five wards will attend Training for Transformation courses, and will receive further training to carry out all the elements in the process approach outlined in Figure 5.2. They will be supported to carry out a process based broadly on the following steps:

- an institutional survey of suitable clubs in their ward
- wealth-ranking exercises
- household and community needs-assessment surveys
- participatory planning meetings
- surveys of local agricultural practices
- exposure visits to view new technologies
- planning, testing and adapting new technologies; and
- linking up with other service providers.

The *Participatory Extension Approaches* manual prepared by ITDG and GTZ (see below) is being used to train Agritex staff to implement these projects.

Preparing an exit from Ward 21

ITDG is aware of the potential danger that its role in Ward 21 might never diminish, as further needs are identified by the community, requiring the facilitation of a process to address them. ITDG's involvement, which is now largely at the request of the community, will continue to decrease, and careful attention is being given to the future sustainability of any present action; interestingly, the 1996 evaluation report commented that the work in Ward 21 was still at a critical stage requiring ITDG's continued support (Scoones and Hakutangwi 1996b). However, project staff are keen to continue to develop confidence among community members, particularly in the latter's relationships with service providers, to whom they increasingly turn for solutions to their problems. It is in this context that a formally constituted CBO in Ward 21 is seen as very important for continuing new initiatives within the ward independently of ITDG (see Box 12.1). At the same time, ITDG will continue to discuss with community members the need for its presence, until it is clear that the time for complete withdrawal has come.

Outside Chivi District – Nyanga District, Manicaland

A second project is currently being developed in Nyanga District, Manicaland Province, based on the approach developed in Chivi. This, it is anticipated, will help to consolidate the evidence for the participatory approach in Zimbabwe, and provide new lessons and experiences for ITDG to learn from, in particular as the geographic, social, cultural and administrative contexts are different from those in Chivi. The Manicaland provincial authorities encouraged ITDG to work in Nyanga District where there is little NGO activity at present, keen interest in the approach on the part of Nyanga District Agritex staff, and other ITDG programme activities already in existence (Lewis 1995).

The Nyanga project was initiated in 1996, when two wards were chosen, Nyanga North Ward 2 and Nyanga West Ward 7. Wealth-ranking and needs-assessment exercises were carried out, and community leaders were sent on the Training for Transformation course. There are a number of differences between this area and Chivi that have affected how the project is operating.

First, there were no pre-existing farmers' groups and gardening groups, so clubs had to be set up at the start. These are based around kraals, but differ from the traditional structure of the kraal since the club leader is elected, and the kraal head participates as an ordinary member of the club. This presented a challenge: how to enable women to take an active part within these traditional institutions, when there were no gardening clubs catering only for them. Surprisingly, however, women in Nyanga had already achieved some prominence, and were represented at the district as well as ward level; this is in contrast to Chivi, where men had been the only office holders at any level.

Second, the Nyanga project has a broader focus than the Chivi project where initial discussions on needs focused on food security and food production constraints. In Nyanga the debate encompassed the full range of constraints faced by the community; for example, young people who were supplementing their farming income from gold panning were aware that because they were not registered as small-scale producers, they were mining illegally, and were there-fore unable to get a good price for their gold from the dealers. By getting together in planning workshops and with the help of the ITDG project staff, they managed to acquire information on the current prices for gold, markets and registration procedures and were able to contact people at provincial officers of the Mining Commissioner. Together, 12 men from three kraals raised over Z$900 and have sent representatives to register their claim.

Box 12.2 The 'discomfort model' of service provision

Many of the kraals in Nyanga are very remote and inaccessible to motorized transport; indeed, one of the first joint project achievements of the farmers of Jimu and Chibvembe kraals was to mend the road leading to the kraal. In August 1998 the first-ever seed fair was hosted in Kudzanayi village, attended by 135 people from the kraal clubs as well as the district administrator and the ward and district agricultural extension officer.

The farmers were eager to take this opportunity to ask the agricultural extension officer for help with several matters, especially the need for pegging contours. Currently most fields in the area are ridden with gullies as there has been no conservation work in the fields. The farmers were aware that without contour ploughing they risked surface run-off and soil erosion during heavy rains, but in order to do this they needed to have the contours measured, and up to this time none of the ward had been marked out because there was a shortage of levelling machines.

The extension officer was unable to meet their request, so the project introduced a simply constructed piece of equipment, the 'A' frame, to mark out the contours (see Figure 8.4). One-day training sessions were held in six sites, and 89 people were trained. Farmers and extension staff were amazed at the simplicity of the technology and the ease with which it could be made and used.

We have seen him or heard about him [the ward extension officer] at some other forum, but this is the first time he has come this way.
Farmers from Chibvembe and Jimu Kraals

I never thought I could stand up and speak in front of a large crowd, let alone in the presence of the district administrator and agricultural extension officer. I feel great about it and I will find it much easier next time.
Mr Musambizi, chairperson of the seed fair committee

What I have seen about the 'A' frame is that it is more thorough than the Agritex department's machine because the pegs are placed every metre of the contour, whereas the machine jumps in some places.
Zekias Ngirishi, farmer

If I had known of this technology earlier on I would have finished pegging the whole ward ages ago. It would be of great help if the project were also to train all the agricultural extension workers of Nyanga North in the use of the 'A' frame.
Mr Mahwaeni, the agricultural extension officer

Another area of concern related to the poor prices being offered locally for the tobacco grown by a number of farmers on the banks of the Nyangombe River. At the Training for Transformation courses the farmers met people from other parts of the country who informed them of other tobacco marketing channels. Now they sell their tobacco to the industrial area in Msasa, where they get a better price.

Some of the villages and kraals in Nyanga Ward 2 are very remote, and face greater problems of inaccessibility than in Chivi Ward 21. This makes it all the more important to build people's confidence to demand help from the agricultural extension officer, who otherwise is unlikely to visit them (see Box 12.2). At the same time, farmers must be encouraged to make the most of their own resources. During 1998 Agritex suffered another budget cut in a series of reductions from the government, and this has affected the mileage allowance of agricultural extension officers. Clearly their ability to visit remote smallholder farmers is limited, which means that farmers will have to rely on their own ingenuity in performing experiments and in passing technologies on to each other.

The work in Nyanga will, it is hoped, also contribute to the debate about participatory approaches not only within Manicaland Province Agritex, but also, through provincial staff, to the national level. This, it was hoped, would increase the influence of the project at national level, and also provide a broader base on which to make the case for participatory development outside Zimbabwe.

Working with national institutions

As well as starting similar projects elsewhere in Zimbabwe, ITDG has used the Chivi project to promote the participatory approach to other organizations. Project and other ITDG staff attend conferences and seminars, reports are shared and articles are written, which use the case of Chivi as an example of this approach in practice. In particular, ITDG has worked closely with two other NGOs based in Masvingo Province: the GTZ–Contill project mentioned above, and IRDEP (formerly CARD). Although the approach of these two NGOs differs slightly from that of ITDG, some common ground has been identified, and the three organizations are working together to promote participatory approaches to agricultural development (both research and extension) within the province.

The participatory approach is gaining a wider audience as project participants are elected to positions of influence in national institutions. In 1997 Mr Z. Gunge, Ward 21 Farmer Club Area Committee Chairman, was elected Secretary to the Chivi District ZFU Committee.

Also in 1997, Mrs L. Chiza, Ward 21 Garden Groups Area Committee Chairperson, was elected to the post of National Secretary for ZFU Women's Wing. This appointment ensures that the voice of a grassroots participant and strong advocate of the participatory approach will be heard at national level.

Figure 12.1   The project is now operating in Masvingo Province and Manicaland Province

Outside Zimbabwe

There have been many visitors, from both inside and outside Zimbabwe, to the project over the last few years. These visits are sometimes mediated initially by project staff, who may be the first point of contact, but who then hand over the details to the local community (usually now through the area committee). The latter then organize the visit, and decide which sites will be visited, who will provide refreshments and so on. Visits have been made to, or received from, the University of Natal and the University of the North in South Africa, the Women's

Resource Centre in Swaziland, the Farming Systems Research Unit from Namibia, the Caribbean (senior agricultural delegates), and a group of farmers, representatives from Ward 21, have visited a project in Tanzania. There is an opportunity to make closer links with government institutions in Malawi, with the support of World Bank and IFAD.

Community voices

In addition to these activities, the project has also encouraged the community to develop ways of sharing their experiences more widely. A workshop was held in early 1994 to discuss what messages they would like to share, and with whom. The participants listed as their target audience: farmers in similar geographical areas, service providers such as Agritex and research departments, other government bodies, and donors. As a result of this planning, they have made two community radio programmes (with the help of ITDG's communications staff), received many visitors and made visits to surrounding areas, and produced, with ITDG, a calendar for sale, in which they describe their activities for each month, illustrated by photographs which they commissioned. As mentioned earlier, a consultant also visited the area in early 1996 to gather stories and experiences of the community, which will be compiled in the Shona language. There are also plans for a video which will feature community members telling their own experiences of the project.

Influencing agricultural extension policy in Zimbabwe

The context of agricultural extension in Zimbabwe

The project's aim of influencing agricultural extension policy towards a more participative approach was developed in the context of the history of extension in Zimbabwe. Agricultural policy in Zimbabwe was based originally on the 'gospel of the plough' developed by Alvord in the 1930s. Maize was promoted as a main crop, intercropping forbidden, and the removal of trees from fields encouraged. Ploughing was adopted in such a widespread fashion that it is now considered to be 'traditional' practice in Zimbabwe. This policy resulted in high levels of soil erosion, which were combated by enforced mechanical conservation works. Consequently, traditional and existing practices among communal area farmers were all but eradicated and farmers' knowledge and skills downgraded (Hagmann, Chuma and Murwira 1995; ITDG 1995a; Mutimba 1994).

The Zimbabwe Government's agricultural extension service, Agritex, is based on a network of ward-level extension workers, supported by a district, provincial and national hierarchy. Extension messages are based on the findings of trials at the research stations, and farmers are rarely encouraged to innovate, experiment or adapt the recommended techniques. Cash crops and improved

Box 12.3   Networking and influence

The project has developed an extensive network of strategic alliances. These include NGOs committeed to promoting participatory approaches: VECO, SAFIRE, the Food and Agriculture Organization-Farmesa project, HIVOS, GTZ food security project, Friedrich Ebert Stiftung, the International Fund for Agricultural Development, IRDEP, and the Institute for Environment Studies. The project works with these organizations to refine and promote wider use of participatory approaches.

The project maintains and widens its network through sharing project documents, participating in workshops, meetings, seminars and exchange visits. In 1996-7 the project team members:

- attended a workshop on drought management in Masvingo;
- attended a workshop on enhancing the impact of research through improved seed supply in Harare;
- participated in the development of the Chivi District local government three-year rolling development plan;
- took part in an African region preparatory meeting on the revision of the International Undertaking on Plant Genetic Resources in Ethiopia;
- attended a training workshop for trainers of participatory technology approaches;
- contributed to ITDG's international food production strategy.

Farmer representatives from Ward 21 and Ward 4 participated in a national workshop on Land and Communities, reviewing recommendations by the Land Tenure Commission. Coopibo invited Ward 21 farmers to visit Mudzi District to share their experiences with farmers they are working with.

varieties which require high levels of inputs (fertilizer and pesticide) usually form the core of these extension messages, which are seldom altered to suit different environments or the differing socio-economic status of farmers. As described above, the women's vegetable gardens are frequently completely overlooked by extensionists, who do not recognize vegetable production as a major agricultural activity making a considerable contribution to household food security (Murwira 1991b).

The key tool for agricultural extension is the master farmer scheme, which involves a two-year training course (of monthly lessons) at the end of which the farmer receives a certificate, if the relevant standards have been met in the farmer's fields. An advanced master farmer certificate is also available, following a further two years' course and a written examination (ibid.). As a result of the focus of the extension worker's time on master farmers, most community members do not get access to the worker: for example, a study in Gutu District in Masvingo Province estimated that extension workers reached on average only 300 out of the 1000 households in their ward.

The findings of government research projects, on which the extension messages are based, are passed up through the chain of command from the research station to the head of the Department of Research and Specialist Services, and thence to the Agritex chiefs at national level who then pass it down their own chain of command, eventually reaching the extension worker. This means that information is often out of date, and some research results never actually reach the farmer. The result of all this, as noted above, is low uptake of extension messages, in particular among the more marginal communal farmers: for example, one survey reported that only five out of 15 recommended technologies were taken up by farmers (Mutimba 1994).

A strategy for influence

In attempting to influence government extension policy, the project's aim is to promote an alternative method of extension, based on the following key aspects:

- Increased emphasis on direct contact and dialogue between farmers and researchers;
- A more equal relationship between farmers and extension workers, with greater respect for farmers' knowledge;
- Decreased emphasis on master farmer training schemes (which tend to exclude more marginal farmers);
- Involvement of government structures to promote sustainable changes in extension approaches (ITDG 1995a).

A strategy for influencing extension policy was drawn up by the project, which involved working at a number of different levels (see ITDG 1993d and Watson 1992 for further details of this strategy). The project worked closely with the extension worker based in Ward 21 from the outset (although this process was somewhat hampered by a change of staff early on). This involved the extension worker's joining planning workshops and community meetings, attending a Training for Transformation course, participating in training and feedback workshops, and so on. The purpose of this was twofold: to share the approach with the extension worker on the ground as it was developed, and to provide evidence for other levels of Agritex that such an approach could be adopted more widely by extension staff.

At district level, Agritex officers were kept informed of the aims and progress of the project through regular reporting by project staff at the District Development Committee, through the sharing of written reports, and through encouragement to visit the project whenever possible.

The provincial Agritex officers, based in Masvingo town, were also kept informed through verbal and written reports and were encouraged to visit the project. Whenever possible, ITDG facilitated opportunities for discussions between them and Chivi farmers.

There was at first no direct contact with national-level Agritex officers; the project planned that they would be best reached through the provincial level. In early 1994, ITDG also commissioned a study of agricultural extension in Zimbabwe, in order to increase its understanding of the policies and practices of the service and inform the influence strategy (Mutimba 1994).

As with local institution strengthening, Training for Transformation courses were a key instrument in influencing extension staff. In early 1993, seven extension workers were trained, including those based in Ward 21 and Ward 4. The response to this training was very positive, both from the extension workers and from the communities to which they returned, to the extent that in early 1994 the district Agritex officers requested funding from GTZ to send all of the district's extension workers for such training. More recently transformative training is also being incorporated into the curriculum for training all extension officers (see next section).

The impact of influencing activities

Progress was slow at first, as the project needed time to develop the approach and to be able to produce evidence that it was an effective way of working with the community. However, the persistent sharing of information with district and provincial level Agritex officers eventually bore fruit.

At district level, Agritex officials in Chivi are adopting a more participatory approach. This is partly in response to direct involvement with the project, through visits and the reporting system described above, which revealed, for example, the widespread adoption in Ward 21 of techniques such as tied ridges which had not been taken up in the past. District-level interest in the approach also grew in response to pressure from the Masvingo provincial officers, who had begun citing Chivi as an example of PTD in national fora.

Chivi District Agritex staff organized several workshops to review their extension strategy. One held in April 1994 included all extension staff in the district, councillors and one farmer representative from each ward. The farmer representative from Ward 21 assisted in the facilitation of the meeting, which focused on farmers' priorities, the constraints of only working with master farmers, and the imposition of extension messages on farmers (Watson 1994).

As a result of this, and other meetings at district and provincial level, Chivi District Agritex has developed a plan for testing and adopting a participatory extension approach, supported by ITDG and the Chivi project. Although ITDG's original plan was that the work in Ward 4 should be the key vehicle for Agritex taking on and implementing the approach, this turned out to be inappropriate, as neither ITDG staff nor Chivi District extension staff were ready at that stage. The new plan was developed by the Chivi District authorities themselves, and thus is more appropriate than one of ITDG's design, in terms of acceptance and owner-ship of the process. The Agritex supervisory areas in which Wards 21 and 4 are located will be the basis for testing out the approach on a wider scale. The exten-sion workers in those areas, and their supervisors, will be trained and supported in implementing a participatory approach, based on the key elements of the Chivi experience (see chapter 5). Feedback workshops spaced at regular intervals will support the extension workers through the process of planning each stage (Murwira 1996).

ITDG's role focuses on training and support, and in distilling the lessons and implications for the roles of supervisory extension officers and other officials, but does not involve direct contact with the communities in these supervisory areas.

At the provincial level, an Agritex all-staff conference was held in April 1995, to which ITDG was invited (together with the GTZ–Contill project) to share the experiences of the Chivi project (staff were encouraged to visit the project before-hand). The conference made the following resolutions:

- To initiate the participatory approach in all seven districts in the province;
- Training for Transformation and participatory rural appraisal techniques to become tools for changing extension staff attitudes;
- The Chivi project and others to be used as models for the province, and as sources of support for the districts;
- Master farmer training to be reviewed and redesigned, to incorporate more marginal farmers, and to become more relevant (ITDG 1995b).

As part of these plans, Masvingo Province Agritex staff organized a training workshop in February 1996 for selected extension workers from all the districts in the province, who would pilot the participatory approach in their area. ITDG, GTZ–Contill and IRDEP all had inputs to this workshop. The provincial training branch is also in the process of selecting a number of extension staff from the province to receive the full Training for Transformation course to enable them to become trainers themselves, as a resource for the province.

Having succeeded to a large extent in gaining acceptance by Agritex (at least at provincial level) of the participatory approach in theory, the project is faced with the challenge of fostering the implementation of this approach in practice. In addition to the plans for Chivi District described above, ITDG, together with Agritex and GTZ, has written a manual called *Participatory Extension Approaches*, and this has been tested on 70 agricultural extension workers in districts within Masvingo Province. The manual draws on experiences from:

- the Chivi Food Security Project
- the Contill project based at Makohili Research Station and projects in Gutu, Zaka and Chivi Districts of Masvingo Province
- IDREP projects based in Zaka District

and is based on the key principles and the key steps of the process shown in Figure 5.2 in chapter 5. These steps may be distilled into the following elements:

- understanding local institutions
- understanding the needs and situation of the more marginal in the community
- building on farmers' knowledge; and
- enabling the community to control the process (ITDG 1995b).

The manual is now being revised to take into account comments that came back from the extension staff testing it. ITDG, Agritex and GTZ and will also be producing a training video of the key elements of the approach, to be used as a tool for in-service training of extension workers.

The approach developed by the project is also being discussed at national level, with Masvingo Province Agritex officials leading the debate, based on the Chivi experience. Agritex's chief training officer jointly carried out a full evaluation of the project in 1996. During a visit of 30 senior agricultural delegates from the Caribbean to Zimbabwe, national Agritex staff directed the visitors to Chivi, to investigate the participatory approach, and in particular the plans for a more participatory master farmer training scheme (ibid.).

As a result of this influence, Agritex has decided to train all its staff in participatory methods over the next five years, and will be using the manual *Participatory Extension Approaches* to do this. The training tries to provide an understanding of the process steps used by both ITDG and the GTZ-Contill project in Masvingo. Agritex have also realized the need for all their staff to receive awareness-raising training first to enable them to understand the importance of the PEA process. Thus, Training for Transformation has become a prerequisite to the participative extension training for all Agritex extension staff. A number of staff in Agritex are now PEA/Training for Transformation resource persons, and they are facilitating training in all the eight provinces of Zimbabwe. All Agritex extension staff, who number more than 2000 and who are certificate holders in agriculture, are being expected to undergo a one-year upgrading diploma course whose main curriculum will be based on Training for Transformation and the PEA process. Both ITDG and GTZ are being consulted for developing the curriculum. Recently GTZ evaluated the impact of this training on Agritex staff and, although the details have not yet been published, preliminary results are positive.

Putting these approaches into practice, Agritex has recently launched another programme in two regions within Masvingo and Manicaland. Both extension workers and community representatives visited Ward 21 in Chivi to learn from the community there. GTZ and ITDG are providing assistance on request to the extension workers implementing this programme.

If any further indication is needed of the seriousness Agritex attaches to these new approaches, it is significant that Agritex recently carried out a review of its performance with respect to smallholder farmers. This revealed that extension workers still need to change their attitudes and ways of working with them, including the need to become facilitators rather than instructors. There are already positive signs, however: in most areas farmers, researchers and extension workers are forming new partnerships in which farmers' experimentation is supported. Extension workers are also facilitating farmers' innovations by taking them to visit other farmers.

13
THE THEORY BEHIND THE PRACTICE

This chapter aims to set the Chivi project in a theoretical context, by outlining the conceptual model on which it is based and the elements of the model which have been shown to be essential. By reflecting on these critical elements it may be possible to identify what is necessary for a food security project to be successful elsewhere: not by prescribing certain project activities, but by isolating certain underlying project principles. The Chivi project does not claim to break any new ground in terms of development theory; what it does is provide a useful (and possibly rare) illustration of how current theory on participative development can be put into practice, and the attendant challenges.

A conceptual model for the Chivi project

The conceptual model for the Chivi project is based on the project's overall aim of enhancing food security, through support to farmer-first sustainable food production (see Figure 13.1 below). Hinchcliffe's statement on sustainable agriculture and food security, prepared for the World Food Summit in November 1996, suggests two factors necessary for this approach: 'Regenerative and low-input . . . agriculture can be highly productive, provided farmers participate fully in all stages of technology development and extension' (Hinchcliffe 1996). These key factors of *participation in technology development*, and *participation in extension*, are taken as the basis for this analysis of the project.

The remainder of this chapter is divided into two parts. Part A discusses the issue of participation in technology development, and in particular the PTD model, in the light of the project's first three principles:

- Building on local knowledge and skills is fundamental for PTD to enhance local food security.
- Participation in decision making for increasing technical capacity and improving technology choice is fundamental for enhanced local food security.
- Strengthening local institutions is essential to achieve this participation.

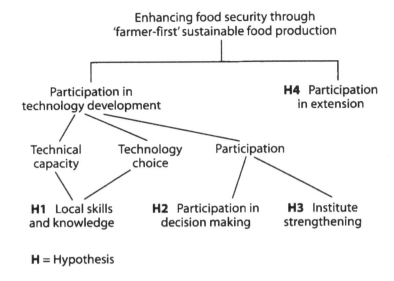

H = Hypothesis

Figure 13.1   Conceptual model for the Chivi Food Security project

Part B considers the issue of participation in extension policy and the project's fourth principle:

● Participation in extension policy is necessary for enhanced food security.

Each part begins with a definition of the key concepts, followed by a discussion of the relevant principles, the activities of the project in support of these principles, the results of these activities, and the issues and challenges which remain.

Part A: Participation in technology development

The debate about participation in development has been going for many years and has included a range of views. In summary, however, there has been a conceptual shift in recent years from 'beneficiaries participation in professionals' projects, to professionals' participation in people's processes' (Farrelly 1995). The debate still rages around the extent to which actual development work is able to achieve this 'real' participation on the ground.

Within the movement for participation in development, there has been a focus on what has been termed participatory technology development (PTD). This concept emerged from a rejection of the standard 'technology transfer' paradigm and seeks to re-evaluate the role of outsiders in technology development in the light of the growing debate on participatory development in general. Technology is defined here as both 'hardware' (tools and equipment) and 'software' (skills, knowledge and training) (Croxton and Appleton 1994). The PTD debate originated in the sustainable agriculture sector, although it has broadened

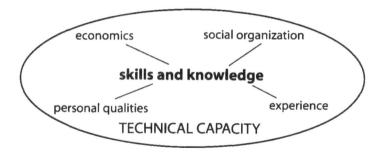

Figure 13.2   Technical capacity

to include other disciplines and is an approach to development, rather than a goal in itself. It consists of three elements: technical capacity, technology choice and participation, all of which have been fundamental to the approach used in the Chivi project.

Technical capacity and technology choice

Definitions

Technical capacity is fundamental to the processes of technology development and change. Croxton and Appleton define this as 'people's ability to identify, adapt, and innovate technology they use or want to use' (1994). However, they also describe it as 'the sum total of the skills and knowledge of a whole community', and also include 'the organizational capacity to manage and use a particular technology' (ibid.), which results in a rather broad definition.

The alternative definition used in this analysis begins with the *skills and knowledge* which individuals in a community possess. These are then constrained or enhanced by a number of other factors:

- personal qualities (for example, innovative ability)
- experience (e.g. education)
- economics (e.g. access to land or other capital)
- social organization (e.g. inter- and intra-household relations and power relationships).

The end product of the relationship between skills and knowledge and the other factors which impinge on them could then be described as technical capacity.

Technology choice is the second element of the PTD model. This is defined here as the range of skills and knowledge made available to people, in addition to that which they already have. An important aspect of technology choice is the need for the user to be able to make an informed choice, thus both the amount and quality of information available to the user is significant. The user also needs to be able to evaluate the results of that choice: this links back to technical capacity and the ability to research and analyse new ideas.

As technology choice is improved, that is, the range of technologies available to users is increased, so their knowledge and skills base is increased and hence their technical capacity is strengthened. In the same way, as Croxton and Appleton point out, if a new idea requires skills which an individual does not have, then technology choice is limited by technical capacity. The two are therefore interdependent.

The first principle

It is clear from the above that skills and knowledge are a central thread running through both the concepts of technical capacity and technology choice. The first hypothesis of the Chivi project's approach is that *building on local skills and knowledge is fundamental for PTD to enhance local food security*.

Project activities in support of the first principle

In support of this principle, the project has focused on a number of key activities. The first, developing and sharing existing local knowledge, involves identifying the skills and knowledge already existing in the community (part of the community's technical capacity), recognizing and giving value to it (increasing their technical capacity) and sharing it with other farmers (increasing technology choice). Second, the exposure visits to other sources of knowledge make an important contribution to improving technology choice. Third, the testing and analysis of new technologies is carried out by farmers themselves, further increasing their technical capacity. Fourth, farmer-to-farmer dissemination aims to improve technology choice for the wider community.

Results of these activities

When asked to analyse the key elements underlying the success of the technical interventions supported by the project, farmers interviewed during the 1996 evaluation claimed that increasing 'awareness and knowledge' were fundamental to this process. Government officials have been surprised at the take-up of some technologies (for example, tied ridges) which they have attempted to promote for some time with little success. This reflects the difference in the project's approach: community members have selected an option for themselves, tested it in their own fields and shared the results with others, thereby increasing both their technical capacity and the range of technologies available to them. Indeed, the 1996 evaluation concluded that the multiplicity of sources of technology information was a positive factor in the project (Scoones and Hakutangwi 1996b: 11). This is because it built on the knowledge and skills which farmers already have, by encouraging them to select technologies which were appropriate to them, and to test, adapt and adopt them in their own fields.

Issues and challenges remaining

Although the results of the project thus far support the principle that building on local skills and knowledge is fundamental for PTD for food security, a number of issues remain:

- The 1996 evaluation report highlighted the challenge facing the project in the future regarding the availability of new information, if all the 'off the peg' solutions at research stations and other sources have been accessed. However, there is considerable potential, as the report points out, to develop further the farmers' involvement in research, in collaboration with the research stations. This presents exciting opportunities for farmers to be more pro-active in defining the extent and direction of the increases in technology choice to be made available to them, as well as further tapping their own knowledge and skills to share with the research community.
- The evaluation report also questioned the sustainability of the exposure visits, once ITDG funds are no longer available to support this activity. However, the process which is now being developed in Ward 4 in Chivi District, is based more on farmer-to-farmer exposure visits (several to Ward 21), which has the benefit of increased sustainability, although it may be more restrictive in the choice of technology options made available to the community.

Participation

Definition

Participation is the third element of the PTD model, which interacts with the elements of technical capacity and technology choice. A useful definition of participation is given by Oakley: 'the process that seeks to increase the influence of disadvantaged groups and so gain access over technologies and other resources that would help to sustain or improve their standard of living' (in Croxton and Appleton 1994). Within the PTD model, participation is the approach used to increase technical capacity and improve technology choice. This means in practice that the users themselves are in the forefront of the process and that the 'outsiders', as Croxton and Appleton call them, play the role of facilitators and assistants, rather than determinants of the process.

Second and third principles

The concept of participation is central to the approach developed in the Chivi project, and forms the second principle, namely that *participation in decision making about increasing technical capacity and improving technology choice, is fundamental for enhanced local food security.*

The third principle, linked to the second, is that *strengthening local institutions is essential to achieve this participation* and for the sustainability of any benefits. In this context, 'strengthening' could be defined as increasing an institution's representativeness (with regard to all sectors of the community, in particular the marginalized), the democracy of its decision making, and its ability to analyse and articulate the concerns of its members.

Project activities in support of the second and third principle

The project has sought to foster participation in decision making and to strengthen local institutions in a number of ways, described earlier. They include:

- the selection of existing local institutions (in particular farmers' clubs and garden groups) as the main partners for the project;
- the provision of Training for Transformation courses to enhance participation at all levels;
- a system of community decision making and feedback meetings.

Results of these activities

These activities have resulted in high levels of community participation in the project. As mentioned above, Farrelly concluded that about 90 per cent of the population of Ward 21 were involved in project activities (1995). The number and size of the farmers' clubs and garden groups has increased during the life of the project, reflecting increased interest and activity (ibid.).

In addition to widespread participation in project activities, there is also involvement in the direction of those activities: for example, Farrelly records that participation in leadership of the farmers' clubs and garden groups is no longer confined to the élite (in particular master farmers).

The 1996 evaluation report concludes that participation has developed institutional capacity among farmer groups in Ward 21 in the following ways:

- an increase in confidence, and the ability to make demands on local service providers, notably the Agritex extension worker;
- an increased network of contacts facilitated by the project, which farmers are now making use of independently of ITDG;
- group cohesion that encourages joint work and problem solving;
- increased capacities of groups to plan, to seek solutions to problems and to manage change

(Scoones and Hakutangwi 1996b: 21).

The government officials interviewed by the evaluation team attributed much of the project's success to the Training for Transformation courses and the 'conscientization' approach (ibid.: 16). The evaluation report also notes the increased networks of institutions in which local farmers are now participating, reporting that 'in the pre–1990 period, an average of 3.3 institutions were linked (some not very closely) [to farmers' food and livelihood security needs] ... in the period since the project started, the average number mentioned increased to 6.2' (ibid.: 17).

The evaluation report describes the project's underlying principle of institutional strengthening as the assumption that 'institutional change (software aspects) is a basic pre-condition to technology development (hardware aspects)' (ibid.: 4), and concludes that the evidence supports this assumption (ibid.: 23). Farrelly similarly concludes that his hypothesis – that the participative approach is necessary to the success of the project – is proven, although he adds several other conditions which are necessary (the availability of Training for Transformation courses and training, partners and funding to pursue technical options). Furthermore, farmers interviewed during the evaluation cited 'working together' as one of the three key elements of the success of the project, thus placing participation and the fruits of institution strengthening at the centre of the project's approach.

Issues and challenges remaining

The evidence of the project thus far supports the second and third principles to the extent that participation in decision making and institution strengthening are fundamental to technology development. However, a number of issues and challenges still remain.

First, the approach taken by the Chivi project is based on the assumption that local institutions are willing and able to change. The project began at a time when the farmers' clubs and garden groups in Ward 21 were at the stage of development which enabled transforming influences to take hold. There also existed sufficient levels of organization within the institutions to provide a base from which the groups could be strengthened. Without these conditions, it is unlikely that the project would have made such progress in institution building and the facilitation of participation.

Second, there is a potential paradox in the dependence of the project on local institutions as the vehicle for technology development, when, as the evaluation report points out, many of the technical innovations which have been identified through the project process have been developed by individuals, not groups. However, this may be addressed by local groups identifying, encouraging and fostering innovation both among their members and other individuals.

Third, there has been considerable dependence on the Silveira House Training Centre in Harare to provide the Training for Transformation inputs which have played a vital role in developing and strengthening individual and institutional capacity in Ward 21. The absence of such a facility would inevitably hinder the replicability of the project's approach in other areas. The costs of such training, which have been met thus far by ITDG (and to some extent by Silveira House), may also prohibit the use of this facility in the future as ITDG withdraws from the project area. However, within Ward 21, and to a lesser extent across Chivi District, there are now a number of potential trainers who can and do share what they have learned from the Training for Transformation courses with other farmers, through workshops, reviews and other meetings.

Fourth, the costs of participation can be high, particularly in terms of time commitment. In order to facilitate joint decision making and keep everyone informed and involved, there are numerous meetings, workshops and feedback discussions taking place in Ward 21, all of which consume time and effort for the participants. For members of the area committee, the time commitment is considerable. The evaluation report also points out that the increased networks available to farmers, discussed above, also bring with them a cost: 'in seeking out information, confirming social relationships... dealing with local political ramifications, managing conflicts...' (ibid.: 32). Ultimately, however, these costs of participation can only be determined by the farmers themselves, who will cease to participate when the benefits, as they perceive them, no longer outweigh the costs of participating.

Fifth, the project's approach of developing local institutions to include all community members has met with some degree of success. However, as the evaluation report notes, underlying the apparent harmony there are inevitably power struggles which remain (ibid.: 30). The wealthier élite, master farmers and others, do not willingly relinquish their hold on community decision making. The tensions between traditional leadership (as represented by the *sabhukus* and chiefs) and the modern structures (such as VIDCOs) have not been completely eradicated, in spite of efforts to achieve involvement of all parties in project activities and decision making, and the search continues for ways to resolve them.

Sixth, one of the most challenging issues relating to participation in the project is that of the excluded minority. Recent studies have shown (for example Farrelly 1995; Scoones and Hakutangwi 1996b) that there is a small minority (around 10 per cent of the local population according to Farrelly) who do not participate in the project. In spite of the Training for Transformation courses and other efforts to promote widespread community involvement and inclusion, therefore, there still exist a number of people who do not participate, and some degree of prejudice against them.

This may be in part a result of the project's strategy of dialogue with the whole community, aiming to change levels of participation and involvement by consensus rather than by direct challenge (see chapter 4). The question remains whether the exclusion of a poor minority invalidates the strategy or not. For example, Farrelly quotes Burley who states 'the totally destitute, the "poorest of the poor", the starving and the displaced are not going to have the mental and physical resources to respond' (Farrelly 1995: 9). On the one hand, the majority of participants are relatively poor people themselves (mostly in the middle two wealth ranks) and there is an argument, supported by project staff, that to target the very poor directly would in fact undermine existing community relationships of responsibility and obligation. On the other hand, the common perceptions of non-participants as 'lazy' and 'stupid' presents a strong challenge to the project and its aim of participation.

There are additional challenges in relation to the gender aspects of participation in the project. As detailed above (see chapter 10), women have participated to a large extent in the project, particularly through the garden groups. However, the gender strategy followed has been covert, encouraging women's participation at all levels whilst not overtly challenging the male hierarchy, in particular male dominance in the leadership of the farmers' clubs. The challenge for the project is to develop a strategy which supports and builds on women's existing roles, whilst at the same time tackling women's strategic gender needs. This may involve a greater involvement, for example in leadership in the farmers' clubs (with consequences for time and other costs). Throughout all this, the strategy must be in keeping with the project's overall approach of dialogue and community-led processes.

Any analysis of the project's achievements is inevitably tempered by the unknown impact of other factors affecting the communities in Ward 21 in the early 1990s. For example, in 1991–2, Zimbabwe suffered the most devastating drought for years, resulting in enormous cattle losses, particularly in dry areas such as Chivi; and in 1991, the Economic Structural Adjustment Programme was initiated, resulting, among other things, in reduced remittances from waged relatives to rural communities and increased prices. The impact of these factors on the project's activities cannot be measured.

Part B: Participation in extension policy

Definitions

While PTD presents a useful approach for developing ways of working with communities to enhance their food security at local level, the effects of such activities are inevitably limited by the policy environment in which they take place. As outlined in its International Food Production Strategy (ITDG 1992c), ITDG's strategic objectives include both 'increasing technology choice for food security' and 'influencing policy'. This commitment to a dual objective is echoed by Hinchcliffe's statement on the productivity of low-input agriculture cited at the beginning of this chapter : 'provided farmers participate fully in all stages of technology development and extension'. Participation in extension policy is defined here as creating an extension service which is client-led, in which farmers set the extension agenda by prioritizing their own needs and directing the activities of the service to respond to those needs.

The fourth principle

The original project objectives include, in addition to those of enhancing local food security and strengthening local institutions, the aim of influencing agricultural extension policy to be more responsive to the needs of poor farmers (see chapter 2). This reflects the project's fourth principle, namely that *farmers' participation in extension policy is necessary for enhanced food security.*

Project activities in support of the fourth principle

In order to achieve this objective of influencing extension policy to include greater participation by farmers, the project has undertaken a number of activities (described in chapter 12). The main focus of these activities has been to encourage the adoption of a farmer-led approach within Agritex, initially at local and district level, and ultimately at provincial and national level. Agritex staff have participated in project planning and implementation; all new government agricultural extension staff now take a Training for Transformation course, and existing staff are also undergoing a one-year diploma course; and the *Participatory Extension Approaches* manual has been adopted.

Results of these activities

These activities, coupled with similar efforts by two other NGOs working in the Chivi area, have resulted in widespread adoption of the principles of a farmer-led extension approach in Chivi District and Masvingo Province Agritex, and with great progress also at national level Agritex (see chapter 12). The 1996 evaluation report attributes this success to three key elements: 'systematic training in Training for Transformation, starting with farmers and moving up organizational hierarchies ...; strengthening of local farmer organizations ...; [and] exposure of farmers and extension workers alike to a range of technologies' (Scoones and Hakutangwi 1996b: 36). The evaluators go on to suggest that the project further develop this 'discomfort model' of influencing extension policy, starting with training and supporting farmers to place demands on their extension worker, who in turn makes demands on his or her superiors in the hierarchy, and so on (ibid.: 38). The evaluation report notes one of the project's assumptions, that is, that 'strong local farmers' organizations are essential to the development of a demand-led agricultural development approach' (ibid.: 29) and presents evidence in support of this assumption.

Issues and challenges remaining

The experience of the project thus far has shown that farmers can be and are involved in creating a demand-led extension service. The activities of the project in this area are, however, based on the assumption that the farmer-led approach, as piloted in the project, is replicable by other institutions such as Agritex and the Zimbabwe Farmers' Union (ZFU). This raises a number of issues.

First, does Agritex (and to a lesser extent the ZFU) have the capacity to take on this model of extension, encompassing as it does a radical change in the way that extension is carried out in the field? This issue of capacity involves a number of considerations: first, skills (for example, facilitation skills and investigatory methodologies, such as those used in the needs-assessment and institutions studies); second, organization (including methods to support a more flexible, interactive and responsive service); and finally, finance (for example, access to transport).

Furthermore, there is a danger that the attempt to institutionalize a flexible approach within a national organization such as Agritex may lead to the promotion of a blueprint for participatory extension, resulting in an inherent contradiction. Signs of this have been seen in the extension service's adoption of the Training for Transformation course material for all their staff. This training commitment, while a useful beginning, is not enough in itself to effect a major change in the way the extension service operates.

Further challenges in scaling up from the pilot project in Ward 21 include: ways of integrating the principles of Training for Transformation community structures (the evaluators suggest using the ZFU structure to achieve this) and through this process strengthening the ability of the ZFU to be able to represent farmers at a higher level more effectively; facilitating the replication of exposure visits at low cost; and maintaining the learning process and lesson sharing.

The importance of documenting the process undergone in Chivi has been noted a number of times, not least by Farrelly, who cites it as a necessary condition of the success of the project. In order to continue to influence Agritex and the wider debate on agricultural extension policy, adequate documentation of the experiences, lessons learned and progress made, is required. This document forms part of that process, along with many of the references cited here. In addition, the new monitoring system (see chapter 11) should provide vital information to contribute to this process.

Finally, the project aims to influence extension policy but does not address the other aspects of agricultural policy which have an enormous impact on farmers in Chivi and similar areas, for example land reform, credit and water development. Indeed, increasing landlessness was among the initial list of priority needs identified by the Ward 21 community at the beginning of the project, but has not been directly addressed, although activities to maximize production and increase reliability of production on small landholdings mitigate this problem in a small way.

Conclusions

Part A above presents some of the evidence to support the project's first three principles, that building on local skills and knowledge; participation in decision making regarding technology development; and strengthening local institutions, are necessary to the success of the project. What is less obvious at first, however, is the link between the 'success' of the project and enhanced food security, which is the overall aim.

At the monitoring workshop held in Ward 21 in late 1995, a list of objectives of the project was defined by the community representatives present (ITDG 1996). As listed above (chapter 9), the first three objectives were: to strengthen co-operation; to share skills and knowledge; and to strengthen household food security. During these and other discussions, it was concluded by community representatives that the project was in the process of fulfilling most of the objectives, with the overall conclusion that most participants have attained food security to a large extent. From this analysis, it could be concluded therefore that the three principles presented above are necessary for the success of the project, and that this success is based on achieving the project objectives, of which a key one remains strengthening household food security.

The fourth principle discussed in Part B – *farmers' participation in extension policy is necessary for enhanced food security* – is, however, more complex to prove. The project has had undoubted success in its objective of influencing agricultural extension policy to become more responsive to farmers and it has succeeded in making the case for a different extension approach, which has been piloted in a number of districts in Masvingo Province, and is being introduced as part of the training of all extension workers. What has been proven here is that *participation in extension policy is necessary for the success of the extension service*, rather than for enhancing food security. To what extent the aim of the extension service is to enhance food security is perhaps a separate debate.

However, farmers' participation in extension policy also results in increased capacity on their part, to articulate their concerns and to challenge other service providers, not only Agritex, but also, for example, the research services and local government. In this way, their confidence, both as individuals and as members of local institutions, is enhanced. This in turn increases their technical and other capacity, and helps them build up their own food security in the ways that they consider to be best for themselves and their families.

The final word should go to one such farmer. Mr Haruzivi, whose farming practices were described in chapter 9, is not exactly typical of all the project participants, neither does his single case prove the necessity of any one principle; nevertheless his response to the project is an affirmation of all of the principles that underpinned it. When he had been allocated his landholding in a most unpromising site in this region of poor, communal lands – 'my plot was the last one and next to a hill that is home to a lot of baboons' – he says he almost cried in despair. Since then, through the project, he has observed a number of soil- and water-conservation techniques, has experimented with them and tried out his own adaptations, with astonishing results in terms of increased yields, additional crops and diversity of vegetables. Other households in the village are busy copying his methods. Mr Haruzivi is proud of the results of his own ingenuity and regards his piece of land in a new light: 'Now I would be prepared to kill to protect my land from being taken away'.

Bibliography

Almekinders, C.J.M. and N.P. Louwaars (1999) *Farmers' seed production: New approaches and practices in seed supply*, Intermediate Technology Publications, London.

Appleton, H. (1994) 'Ownership through participation', *Appropriate Technology Journal*, June 1994, Vol. 21, No. 1.

Arrighi, G. and S. Saul (1973) *Essays on the political economy of Africa*, Monthly Review Press, New York and London.

Beach, D.N. (1980) *The Shona and Zimbabwe 900–1850: an outline of Shona history*, Gweru, Mambo Press, Zimbabwe.

Bourdillon, M. (1987) *The Shona peoples: an ethnography of the contemporary Shona with special reference to their religion*, Gweru, Mambo Press, Zimbabwe.

Carney (1999) *Approaches to sustainable livelihoods*, Overseas Development Institute, London.

Chambers, R. (1983) *Rural development: putting the last first*, Longman, Harlow.

Chambers, R., A. Pacey and L. A. Thrupp (eds. 1989) *Farmer First: farmer innovation and agricultural research*, Intermediate Technology Publications, London.

Cheater, A.P. (1989) 'The Problem of Defining Rural Development', Rural Alienation: The Relationship Between Peasantry and Government. Unpublished. University of Zimbabwe.

Croxton, S. and H. Appleton (1994) 'The role of participative approaches in increasing the technical capacity and technology choice of rural communities'. Paper presented to NRI conference, September 1994.

DFID (1999) *Sustainable livelihoods guidance sheets*, DFID, London.

Farrelly, M. (1995) *Whose project is it anyway?: a study of participation in Chivi Food Security Project, Zimbabwe*. Development Administration Group, University of Birmingham, October 1995.

Frischmuth, C. (1997) 'Gender is not a sensitive issue: institutionalizing a gender-oriented participatory approach', IIED Gatekeeper Series No. 72.

Gaidzanwa (1988) 'Women's land rights in Zimbabwe: an overview', RUP Occasional Paper No. 13, Department of Rural and Urban Planning, University of Zimbabwe, Harare.

Hagmann, J., E. Chuma and K. Murwira (1995) 'Kutaraya (let's try!): reviving farmers' knowledge and confidence through experimentation'.

Hagmann, J., E. Chuma, K. Murwira, and E. Moyo (1995) 'Transformation of agricultural extension and research toward farmer participation: approach and experiences in Masvingo Province, Zimbabwe.' Discussion paper presented at workshop 'Extension intervention and local strategies in resource management: new perspectives on agricultural innovation in Zimbabwe', January 1995.

Hagmann, J. and K. Murwira (1996) 'Indigenous soil and water conservation in southern Zimbabwe: a study on techniques, historical changes and recent developments under participatory research and extension', IIED, Issue Paper No. 63, London.

Hinchcliffe, F. (1996) *Sustainable agriculture and food security.* Briefing note for the World Food Summit.

Holleman, J.F. (1952) *The Shona customary law, with reference to kinship, marriage, the family and the estate,* New York, Oxford University Press.

ITDG (1991a) *Chivi Food Security Project: Project Document.*

ITDG (1991b) *Summary of findings of a survey carried out in villages E and F of Ward 21: Chivi District, September 16t–29t. Needs assessment and baseline survey report.*

ITDG (1992a) *Chivi Food Security Project, Zimbabwe.* Proposal to Comic Relief, April 1992.

ITDG (1992b) *Chivi Food Security Project Monthly Report,* April 1992.

ITDG (1992c) *International Food Production Strategy.*

ITDG (1992d) *Workshop on fencing of group gardens at Musvovi School,* November 1992.

ITDG (1992e) *Chivi Food Security Project Monthly Report,* December 1992.

ITDG (1992a) *Chivi Food Security Project Monthly Report,* February 1993.

ITDG (1993b) *Annual Project Review 1993: Chivi Food Security Project,* August 1993.

ITDG (1993c) *Chivi Food Security Project Monthly Report,* August 1993.

ITDG, 1993d. *Chivi Food Security Project Monthly Report.* October 1993.

ITDG (1993e) *Strengthening local food production, Masvingo Province, Zimbabwe.* Application to ODA Joint Funding Scheme, November 1993.

ITDG (1993f) *Chivi Food Security Project Monthly Report,* December 1993.

ITDG (1994a) *Chivi Ward 21 Community Review of Food Security Project,* July 1994.

ITDG (1994b) *Annual Project Review 1994: Chivi Food Security Project,* August 1994.

ITDG (1994c) *Chivi Food Security Project Monthly Report,* September 1994.

ITDG (1994d) *Chivi Food Security Project Monthly Report,* October 1994.

ITDG (1994e) *Leaders Workshop: Ward 21,* 16 November 1994.

ITDG (1994f) *Gender Study: ITDG Chivi Project,* December 1994.

ITDG (1995a) *Chivi Food Security Project: Project Document 1995/96–1997/8.*

ITDG (1995b) *Strengthening local food production, Masvingo Province, Zimbabwe.* Annual Report to ODA Joint Funding Scheme, Project 952, June 1995.

ITDG (1995c) *Chivi Food Security Project Monthly Report*, June 1995.

ITDG (1995d) *Workshop held at Musvovi Hall*, 18 October 1995.

ITDG (1996) *Developing a participatory monitoring system*. Document 1: July 1995; Document 2: October–November 1995; Document 3: Workshop on improving monitoring systems within Chivi Food Security Project, 31 January 1996.

Lewis, V. (1995) *Overseas Visit Report*, OVR 105/95, ITDG.

Lewis, V. (1996) *A synthesis of the gender outcomes of food security projects in Sudan, Kenya and Zimbabwe*, ITDG.

Makumbe, J. (1996) *Participatory development: the case of Zimbabwe*, University of Zimbabwe Publications, Harare, Zimbabwe.

Manyame, C.M. (1994) *Impact of interaction between men and women on women and development: culture survey*, Zimbabwe Women's Resource Centre and Network, Harare, Zimbabwe.

Maurya, D.M. (1989) 'The innovative approach of Indian farmers' in Chambers et al. op. cit.

Mbetu, R. (1997) 'Rural development in practice – the process approach: capacity building and empowerment, experience from Zimbabwe', PhD thesis, Coventry University.

Mulvany, P., B. O'Riordan, and H. Wedgwood (1995) 'Taking root...gaining ground: diversity in food production for universal food security', Paper presented to the Development Studies Association Conference, ITDG.

Murwira, K. (1991a) *NGO and government ministries activities in Chivi District, Masvingo*, ITDG.

Murwira, K. (1991b) *Report on institutional survey in Ward 21 (Chomuruvati area) in Chivi District, Masvingo Province, Zimbabwe, Agriculture and Fisheries Sector*, ITDG.

Murwira, K. (1991c) *Report on the present status of farmers' clubs and group gardens in villages and E of Ward 21, Chivi District*, ITDG.

Murwira, K. (1992a) *A report on investigations in both traditional and current practicies on water harvesting and crop pests control in Ward 21 (Chomuruvati area), Chivi District*, ITDG.

Murwira, K. (1992b) *Report of exchange visits to (a) Chiredzi Research Station, Masvingo, (b) Makoholi Research Station, Masvingo, (c) Mutoko Project, Mashonaland East, (d) Fambidzanai Training Centre, Mt Hampden, by farmers representatives from Ward 21, Chivi*, ITDG.

Murwira, K. (1994) 'Community-based sustainable development: the experiences of the Chivi Food Security Project', Paper presented to the IIED conference on Community-based Sustainable Development, July 1994.

Murwira, K. (1995) 'Freedom to change – the Chivi experience', *Waterlines*, April 1995, Vol. 13, No. 4.

Murwira, K. (1996) Personal communications from ten days of interviews, February 1996.

Mutimba, J. (1994) *Agricultural extension policy and practice in Zimbabwe*. Study report for ITDG, February 1994.

Pacey, A. and A. Cullis (1992) *A development dialogue: rainwater harvesting in Turkana*, Intermediate Technology Publications, London.

Page, S.L.J. and H.E. Page (1991) 'Western hegemony over African agriculture in Southern Rhodesia and its continuing threat to food security in independent Zimbabwe', *Agriculture and Human Values*, Fall 1991.

Rhoades, R. (1989) 'The role of farmers in the creation of agricultural technology', in Chambers et al. (ed.) op. cit.

Rodney, W. (1972) *How Europe underdeveloped Africa*, Bogle L'Ouverture Publications, London.

Rukuni, M. and C.K. Eicher (1994) *Zimbabwe's agricultural revolution*, University of Zimbabwe Publications, Harare.

Sanghi, N.K. (1989), 'Changes in the organization of research on dryland agriculture', in Chambers et al. (ed.) op. cit.

Scoones, I. and J. Thompson (1994) Introduction in: Scoones, I. and J. Thompson, (ed.) *Beyond Farmer First: rural people's knowledge, agricultural research and extension practice*, Intermediate Technology Publications, London.

Scoones, I. and M. Hakutangwi (1996a) *Hazards and opportunities*, Zed Press, London.

Scoones, I. and M. Hakutangwi (1996b) *Chivi Food Security Project: evaluation report*.

Vela, M. (1996) Personal communications from two days of interviews, February 1996.

Watson, C. (1992) Trip notes: visit to Zimbabwe August 1992.

Watson, C. (1993a) Trip notes: visit to Zimbabwe 11–23 February 1993.

Watson, C. (1993b) Trip notes: visit to Zimbabwe 7–19 August 1993.

Watson, C. (1994) Trip notes: visit to Zimbabwe April 1994.

Wedgwood, H. (1996a) Food security slide talk notes, ITDG

Wedgwood, H. (1996b) Food Security Briefing Paper.

Wedgwood, H. (1997) Full visit report, Zimbabwe.

Win, E. (1996) *Our community ourselves: a search for food security by Chivi's farmers*, Intermediate Technology, Zimbabwe.

World Bank Operations Evaluation Department (1994) *Agricultural extension: lessons from completed projects*, World Bank, Washington DC.

Appendix 1: Project chronology

1990

- Initial discussions between ITDG staff in Zimbabwe and UK to plan project.
- Two consultancy reports led to selection of Chivi District as an area representative of communal areas in Zimbabwe suffering household food insecurity.
- Project officer recruited.

1991

- Selection of two wards for focus of activities, by District Land-Use sub-committee, according to criteria set by ITDG:
  - representative of communal areas
  - natural regions IV and V
  - subsistence farming area
  - lack of infrastructure
  - lack of NGO activity
  - food insecurity.
- Confirmation of ward selection by ward councillors.
- Survey of government and NGO activities in Chivi District, to gain understanding of the roles of various institutions.
- Introduction of ITDG to Ward 21 by district officials, through a series of ward meetings.
- Institutions study in Ward 21, covering traditional, formal and informal institutions, with a particular focus on those involved in food security.
- Selection of institutions to work with by ITDG: farmers' clubs and garden groups.
- Wealth ranking in two poorest villages ('focus' villages) in Ward 21.
- Needs-assessment and household studies in those villages, using the wealth-ranking data to select sample households. Feedback of results to community meeting, including choice of farmers' clubs and garden groups as institutional basis for the work.
- Planning meeting with representatives of selected institutions, together with community leaders, to prioritize needs and agree on future plans.
- Decision to focus initially on water and pest control. Recognition of the need for co-operation.

1992

- Study of farmers' clubs and garden groups in the focus villages.
- Study of traditional and current practices in soil and water conservation and pest control. Feedback meetings to discuss findings and strengths and weaknesses of these practices.
- Community meetings to select two farmers' clubs and two garden groups (one in each village) to pilot project activities.
- Exposure visits to relevant research stations and institutions by representatives of the pilot groups.
- Feedback of findings from exposure visits at community meetings, followed by selection of techniques and technologies to try out.

- Group representatives began Training for Transformation (leadership training) courses.
- Pilot groups received training in selected technologies from research station staff, and tested them in their fields and gardens. Activities included: sub-surface irrigation; pest management; water harvesting; crop diversification; shallow-well improvement. Mini-reviews carried out on regular basis with pilot groups and other village members to share information on progress and problems.
- Community review held with representatives of selected villages to assess progress and plan activities for the coming year.
- Demonstrations by farmers and research station staff to show results thus far, including modifications to the technologies, to the other groups within the two focus villages.
- Institutions study carried out in remaining villages in Ward 21.

1993

- First group of seven Agritex extension workers went on Training for Transformation course.
- Garden groups decided on plan to fence their gardens and send two representatives to Harare for training in fence making.
- Seed fair held to share knowledge of crop varieties. Two focus villages exhibited, but the whole ward was invited to attend.
- Planning meeting in remaining villages in Ward 21. Priority needs similar to those in the first two villages, with the addition of strengthening institutions; marketing for vegetables; water transport and post-harvest grain management. Demonstrations of selected technologies by farmers within Ward 21.
- Community review with all villages in Ward 21.
- Second project officer recruited.
- Work on other prioritized needs taking place.
- Ward meeting to explore ways of improving co-ordination: decision to elect new members to existing farmers' clubs' area committee to make it more representative.

1994

- Agritex meeting with researchers, farmer representatives from Ward 21 and ITDG to learn about the project's strengths and weaknesses.
- Garden groups set up their own area committee.
- Work on developing and implementing technologies in response to prioritized needs ongoing in all villages in Ward 21. Community review in Ward 21.
- Seed fair and garden groups field day competition held in Ward 21.
- Work began in Ward 4 with initial community meeting to explain ITDG's way of working. This was followed by village awareness-raising meetings, to build the confidence of the community in their ability to plan and implement their own activities. Extension worker more involved from the beginning of the process.
- Institutions study in Ward 4.
- Village planning meetings in Ward 4, to identify needs and explore solutions in each village in the ward.
- Wealth ranking then needs-assessment study in Ward 4.
- Community meeting to discuss results of the needs-assessment and the institutions studies. Agreed to work with farmers' clubs and garden groups as key institutions in Ward 4.

1995

- Agritex workshop for feedback from staff who attended Training for Transformation courses to share ideas on approach.
- Ward 4 planning meeting, to analyse priority needs, discuss possible solutions based on local knowledge, and plan future activities.
- Exploration of other technical solutions through exposure visits to Ward 21.
- Ward 4 community leaders and group representatives on Training for Transformation courses.

- Demonstrations of techniques and technologies in Ward 4 by farmers from Ward 21. Testing of technologies in Ward 4.
- Ward 4 community review.
- Seed fair, garden groups show and community review in Ward 21.
- Ongoing implementation of activities and testing of new technologies in Ward 21.
- Ward 21 area committees for farmers' clubs and garden groups wrote funding proposals for a revolving fund to send to local donor organizations.

1996

- Workshop for selected extension workers to plan the implementation of the process approach by Agritex in Masvingo Province.
- Fourteen agricultural extension workers from Masvingo Province attended phases I and II of the Training for Transformation course so that they were able to become trainers themselves in the province.
- External valuation report produced jointly by Agritex's Chief Training Officer and Ian Scoones, IDS.
- Participatory monitoring system developed by the project and project team.
- Report produced: local knowledge and systems of on-farm seed production and maintenance.
- Thirty-three farmers in Ward 21 hosted on-farm trials in collaboration with Chiredzi Research Station.
- Six participants trained in making and using a fence-making machine.
- Reports submitted to Comic Relief and ODA Joint Funding Scheme.
- Project started in Nyanga District, Manicaland. Wealth-ranking and needs-assessment exercises carried out in Ward 2; Training for Transformation phases I and II carried out for 37 farmers.

1997

- Marketing points set up by private buyers in Wards 21 and 4.
- Community members receive training in welding, tie-dyeing cloth, sewing, soap-making, etc. HIVOS six-monthly report submitted.
- Ward 21 community members attend a one-week course on food-processing, preservation, and storage.
- Funding proposals submitted to DANIDA and DFID/Joint Funding Scheme.
- Garden groups in Ward 4 set up revolving fund for productive or household assets.
- Eighteen Chivi farmers carried out collaborative research trials with Chiredzi Research Station on varieties and fertility.
- Five field days held to share observations and lessons learned from the trials.
- Three farmers hosted varietal trials on sorghum and millet in collaboration with GTZ–SADC food security unit.
- Two hundred and fifty people attend field days to share results of trials.
- Nineteen men and 13 women trained in gender awareness.
- Differentiation study report produced.

Nyanga project

- Ward 2 planning meeting held to prioritize needs and plan future work with relevant support institutions. Traditional and current farming practices surveyed.

1998

- The *Participatory Extension Approaches* manual is produced jointly with Agritex and GTZ, and is tested with 70 agricultural extension workers. An introductory workshop on PEA was held for all Agritex staff in Chivi.
- Training of agricultural extension workers and farmers' leaders in five communities, accompanied by the start of project activities in Wards 1, 10, 23 and 28 of Chivi District and Ward 9 of Mwenezi District.

- Planning session to register Ward 21 as a community-based organization (CBO). Representatives from Ward 21 also received training in skills such as blacksmithing, tie- and dyeing cloth and poultry-keeping, funded by HIVOS.
- Fourteen men and 14 women from Ward 21 visit ICRISAT Matapos Research Station, and conduct on-farm trials of six sorghum and six pearl millet varieties.
- More than 300 farmers and representatives of support institutions attend the seed fair at Musvovi Hall, Chivi District.
- A study to identify innovative farmers in soil- and water-conservation was carried out and documented.

Nyanga project, Ward 2

- The first seed fair held in Kudzanayi village.
- Problem analysis and planning meetings held for 12 out of 16 kraal clubs.
- Two Training for Transformation courses held for 52 farmers.
- Exposure visits to Chivi and Zvishavane for 19 farmers and four Agritex staff; further visits planned. Testing of technologies in interested farmers' fields.
- Gender awareness courses held.
- Surveys of Ward 8, Nyanga District, where the project will be beginning activities.

1999

- Agritex decides to train all its staff in PEA over the next five years using:
  - The manual, *Participatory Extension Approaches*;
  - Agritex resource persons trained in PEA/Training for Transformation, who are conducting transformative training in eight provinces of Zimbabwe.
  - Two thousand Agritex extension workers (certificate holders) will be required to take an additional one-year diploma course with a strong component of PEA and Training for Transformation. The curriculum was still being developed at the time of writing.

Nyanga Ward 2

- Planning with the remaining four kraals in gardening and dryland farming activities.
- Evaluation of field trials of technologies.
- Training of village borehole mechanics.
- Kraal seed fairs take place, followed by a ward seed fair.
- Nyanga Ward 8: work starts.

Chivi Extension Project

- Develop farmer leaders' training programme in PTD.

Appendix 2: Financial assessment 1989–97

The project's financial costs focused on three main areas: the salary and expenses of the project officer (and later additional project staff); the project vehicle; and the costs of exposure visits and training courses. No other grants or donations have been made towards community activities. There has therefore been a high financial investment in the facilitation of the process and very little in capital outlay or infrastructure. The first project officer was joined by a second project officer in late 1993. In 1994, a part-time social scientist and a secretary were also recruited reflecting the expansion of the project into Ward 4.

Project budget

Year	Project-based staff	Budget (Z$)
1989/90	0 (consultancy report)	125 000
1990/91	0	143 750
1991/92	1	125 000
1992/93	1	320 000
1993/94	1	458 000
1994/95	3	731 400
1995/96	4	1 441 762
1996/97	4	1 289 500
TOTAL	14	4 634 412

Analysis

	Zimbabwe dollars	*£ sterling equivalent*
Total cost	4 500 000	345 000.00
Average cost/year (1991–7)	750 000	57 500.00
Total cost per beneficiary household	2 250	172.50
Average annual cost per beneficiary household	375	28.75

Notes

- UK staff salary costs are not fully reflected in this table. A large part of these salary costs were covered by DFID's block grant to ITDG.
- Exchange rates have fluctuated by more than 100 per cent over the life of the project. Budget figures reflect this variation, based on actual rates at the time of draw down.
- Beneficiaries are taken to be those community residents participating in farmers' clubs and garden groups. Out of a total population of 1200 households in Ward 4 and 1300 households in Ward 21 there were 2000 garden group members and 1700 farmers' club members. The cost analysis is based on an estimated total of 2000 beneficiary households in Wards 21 and 4, where each woman garden group member is assumed to represent a household (average household size is estimated at seven).
- The budgets for 1989/90 and 1990/91 were mainly on consultants which included expatriates to be paid in £ sterling. The budget for 1991/92 covered the project officer's salary, vehicle purchase, maintenance, four international visits, local travel costs, workshops and seminars, communication and stationery as well as local overheads. So for the first two years (1989 to early 1991) expenditure on the project was mainly in the UK.

www.ingramcontent.com/pod-product-compliance
Lightning Source LLC
Jackson TN
JSHW071342130125
77033JS00034B/1017